Introduction to Linear, Parametric and Nonlinear Vibrations

Introduction to Linear, Parametric and Nonlinear Vibrations

MATTHEW CARTMELL
B.Sc., Ph.D., C.Eng., M.I.Mech.E.
Lecturer
University of Aberdeen

CHAPMAN AND HALL
LONDON•NEW YORK•TOKYO•MELBOURNE•MADRAS

UK	Chapman and Hall, 11 New Fetter Lane, London EC4P 4EE
USA	Chapman and Hall, 29 West 35th Street, New York NY10001
JAPAN	Chapman and Hall Japan, Thomson Publishing Japan, Hirakawacho Nemoto Building, 7F, 1-7-11 Hirakawa-cho, Chiyoda-ku, Tokyo 102
AUSTRALIA	Chapman and Hall Australia, Thomas Nelson Australia, 480 La Trobe Street, PO Box 4725, Melbourne 3000
INDIA	Chapman and Hall India, R. Sheshadri, 32 Second Main Road, CIT East, Madras 600 035

First edition 1990

© 1990 Matthew Cartmell

Typeset in 10/12 Times by
KEYTEC, Bridport, Dorset
Printed in Great Britain by
T. J. Press (Padstow) Ltd, Padstow, Cornwall

ISBN 0 412 30730 8

British Library Cataloguing in Publication Data
Cartmell, Matthew
Introduction to linear, parametric and nonlinear vibrations.
1. Vibration
I. Title
531.32

ISBN 0-412-30730-8

Library of Congress Cataloging-in-Publication Data
Cartmell, Matthew, 1958–
Introduction to linear, parametric, and nonlinear vibrations/Matthew Cartmell.
p. cm.
Includes bibliographical references.
ISBN 0-412-30730-8
1. Vibration. 2. Vibration, Parametric. I. Title.
TA355.C34 1990
620.3–dc20
89-71225
CIP

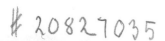

Contents

To the memory of my father

Acknowledgements

I would like to thank Ms Anne Shipley and Mr Stuart Cutts of the University of Aberdeen for their respective efforts in the typing of the manuscript and the production of the illustrations. I have had much useful input from my academic colleagues within the Department of Engineering, and would therefore like to thank Professor Allan D. S. Barr, Dr David Watt, Dr Richard D. Neilson, and also Dr J. W. Roberts of the Department of Mechanical Engineering in the University of Edinburgh for his help and comments. I am also greatly indebted to Dr Watt for his informed and perceptive comments on the manuscript as a whole. I have incorporated acknowledgements within the text where I have leant heavily on the work of others. I would also like to acknowledge the very considerable debt of gratitude that I owe to Raymond G. Fielding for inspiring an enthusiasm for all things mechanical and electrical. Finally I wish to thank my wife Fiona for her interest and support for the project, and for her patience with the many crises of confidence that beset me.

Matthew P. Cartmell

Preface

This book is intended to provide an introduction to some of the interesting, and perhaps surprising, phenomena often encountered in systems which vibrate and which do so under the influence of parametric and/or nonlinear effects. Therefore the bias of the book is predominantly towards the phenomenological, but mainly in the context of mechanical engineering. I have, in places, attempted to bring in applications or circumstances outside mechanical engineering within which particular effects have been observed. However, in the main, the book is directed towards mechanical engineers. There is no shortage of excellent texts in the general area of parametric and nonlinear vibrations but in almost all cases there is an assumption either of prior knowledge and experience or of a very considerable facility with the appropriate forms of analysis. Clearly having the former will only serve to clarify and add to the picture obtained from studying advanced texts, but since many aspiring students in the field will probably not be experienced researchers a more gentle introduction, such as may be found within this book, should be of some help. In addition to this, the second point raised above on analytical skill merits some attention. I hope that by virtue of the reasonably detailed examples given in the book some of the subtleties of certain parametric instability problems, and of aspects of nonlinear vibrations, are brought out without getting too lost in the mathematics. In this context it might be appropriate to draw the reader's attention to certain specific sections within the text and to offer a brief explanation for the length of the discussion at those points. Section 1.7.4 deals with Virtual Work and Lagrange's Equations in the standard manner with minimal abbreviation. This is to allow the interested reader to work through this important analysis and perhaps to gain a clearer understanding of what it means in the process. On the other hand, the result may be noted without going through the derivation; either option is available. A similar approach has been taken in Section 2.1 in which a reasonably concise, yet formal, appraisal of the stability of Mathieu–Hill type equations is presented. There is a brief review of electronic applications of parametric systems in Section 2.4.4(b) and the aim here is to draw the reader's attention to these and to attempt to show that parametric amplifiers are deliberately stabilized by means of specifically introduced nonlinear circuit elements. The discussion is relatively superficial given that this is not mainstream material; references for further reading are given. Sections 3.1.3(a) and (b) treat the kinematics and the derivation of the equations of motion respectively for a parametrically excited cantilever beam where combined bending and torsional motion is possible. Again a reasonably full treatment of the problem is given so that the source of the nonlinearities that are highlighted is made clear. Dealing with a 'simple' problem in this way hopefully goes some way towards showing the potential complexity of nonlinear vibrations without being completely inpenetrable. In the final chapter a section on chaos is included (Section 5.3). I have not attempted to deal with the subject in any great depth but have presented some principals and definitions that I consider important. Several excellent books on chaos have recently been published and the reader is strongly advised to investigate these for further details.

The general intention behind writing this book has been to provide a starting-off point for those who wish to extend their appreciation of mechanical vibrations into the parametric and nonlinear domains. I have provided a list of references which is fairly broad in coverage in order to enable the student to get started, hopefully without being put off by the immensity of the available literature. I therefore strongly recommend that the references to other texts are followed up so that further insight is obtained.

Matthew P. Cartmell
Department of Engineering
King's College
University of Aberdeen

1

Linear vibrations in mechanical engineering

INTRODUCTION

It is intended that this first chapter should serve as a general and broad-based revision of the concepts of mechanical and structural vibrations, specifically those which are conventionally classified as being 'linear'. Although it is assumed that the reader has some familiarity with vibrations theory and appropriate areas of applied dynamics, a substantial awareness of the subject is not a prerequisite for gainfully reading this, or further chapters. It is suggested, however, that appreciation of the central theme of the book, namely parametric and non-linear vibrations, will be enhanced by referring to this section.

1.1 CLASSIFICATION OF VIBRATION PROBLEMS

There are several ways in which we can attempt to classify a system, or a problem in vibrations; for instance it may be conservative, where the total energy of the system remains constant during motion, or non-conservative in which case energy is expended in overcoming some form of dissipation, and where there may also be a kind of forcing or excitation acting on the system. An alternative approach would be to relegate these criteria to a different level of classification and to adopt the terms linear and non-linear as more fundamental categories of problem definition.

Linear vibrating systems will contain mass (or inertia), stiffness, and damping (to some degree) and so, as long as these quantities are themselves linear in their behaviour and are not varying with time, a mathematical model employing a linear ordinary differential equation with constant coefficients should adequately portray the motion of the system. Since damping, or dissipation, is never totally absent in practice we generally accept that a non-conservative linear model will often prove to be a realistic starting point in our explorations. Further assumptions enter into our conceptualization of linear vibrating systems in the form of inelastic mass and inertia elements, massless springs and dampers which possess neither elasticity nor mass but introduce dissipatory forces proportional to the relative velocity across the element. An important and unique identifier of linear action is the principle of superposition which simply states that resultant oscillations may be composed of two, or more, separately excited motions which combine linearly to generate one response motion. The principle can be used to determine the complete solution to a forced vibration problem by combining the particular solution due to the forcing and the complementary solution arising from the initial conditions.

It will probably be apparent, even at this stage, that a truly linear system is unlikely to be found except on paper and that practical questions must surely threaten our convention of linearity, and in particular the notion of easily obtained solutions to manageable and well-behaved equations of motion. Therefore we must be confident that the type of linear model proposed is adequate in the sense that the distributive nature of mass, elasticity and damping, as pertaining to the specific problem, is properly understood. To do this we can additionally categorize the problem as being discrete or continuous in form.

A discretized system may contain one element of mass, connected to attendant massless springs and damping devices, or several lumped masses, or inertias, joined together by massless springs and dampers. A model of this type will rarely be wholly representative of the system under investigation, since it is often difficult to break down real machine and structural problems in the certainty that one is creating an accurate mathematical portrayal. On the other hand, a continuous formulation of the problem, where the properties of mass and elasticity are each considered to be inseparable along the axis of action, will avoid such hazardous simplifications but this will be at the expense of subsequent analytical ease and convenience.

In the case of discrete models one needs to consider the number of degrees of freedom of the problem. This is the number of independent coordinates that are needed to fully describe the possible motion of the system. It does not always follow that each mass or inertia has merely one degree of freedom since its motion may consist of a variety of different translations or rotations, or both, however it does follow that any discrete system will have a finite number of independent coordinates and therefore a finite number of degrees of freedom. A continuous model must by definition have an infinite number of degrees of freedom as its mass is continuously distributed within the structure.

In addition to the above a system may appear to be linear only under certain conditions, and a commonly occurring example of this is the case of small oscillations about the equilibrium point. The requirement that a spring obeys Hooke's law is only adequately satisfied for relatively small deflections, after which its linearity may be in question. Similarly, classical linear viscous damping, as so often called upon in engineering problems, may only hold over a restricted range of velocities, and even then it may not be a realistic indication of that which actually exists within the system.

In conclusion, therefore, it can be said that our interpretation of vibration problems in engineering is of prime importance, and that the answers obtained from analysis will generally involve some form of approximation, so it is crucial for the engineer to develop a logical as well as an instinctive facility with the domains of linear modelling, modelling of nonlinear problems where linearization may under certain conditions be acceptable, and entirely nonlinear problems, where alternative techniques must be applied.

1.1.1 Classification in terms of constraints

Because all problems in dynamics involve formulating relationships between forces as well as amongst displacements it is obvious that forms of constraint must exist, which, in a practical sense, serve to restrict or contain the operation of the system by means of boundary effects or internal interactions of some description. This implies that if we can represent such constraints mathematically then we have a means of specifying relationships between coordinates as well as between forces and torques. This in turn points to the possibility of a distinction between those coordinates that can be related in this way and those that cannot. Coordinates that do not feature in equations of constraint must therefore be independent and will equate to the numbers of degrees of freedom active in the system. These can be translational or rotational, relative or absolute. We can now see that each constraint equation that exists for a system will reduce the number of truly independent coordinates by one (and hence the number of degrees of freedom).

Many cases of structural and machine vibrations allow for the identification of a set of independent or generalized coordinates by considering the constraints acting, and thus the constrained coordinates which result. We use the term holonomic to apply to these. Conversely this process of elimination of constrained coordinates to render a set of generalized coordinates is not always possible, for example we can consider instances in which the active constraint equations contain, say, velocity dependent forces, such as those that give rise to relative sliding velocities in structural joints as a case in point. These constraint equations are unlikely to be integrable, and as a result the system is said to possess non-holonomic constraints.

Within the domain of holonomic constraints two supplementary constraint types are available and these refer directly to the presence of time dependency, so for the case of a time variant holonomic constraint system we use the term rheonomic holonomic. The time-invariant case is called scleronomic holonomic.

By considering and briefly discussing the ideas of conservation, linearity, discrete and continuous distributions of mass and elasticity, and constraints, we have mapped out ways in which we can begin to identify our problem and then we can proceed to devise a useful solution. Application of the tools of classification may not always be a wholly conscious step in the process, however implementation of these ideas to some extent (and perhaps not always explicitly) will always be necessary to derive descriptive equations of motion.

1.2 FUNDAMENTALS OF VIBRATING MECHANICAL SYSTEMS

All mechanical engineering structures which undergo oscillatory motion will contain, and be expressible in terms of, mass, stiffness, and damping. Useful related parameters are inertia (synonymous with mass but found in rotating systems), flexibility which can be shown to be the inverse of stiffness, and dissipatory forces which embrace 'damping' but also describe other non-conservative force effects such as dry friction.

The way in which we choose to identify the significant effects of these mechanical parameters is of great importance since their effect can be straightforwardly direct, or alternatively seemingly indirect in the case of effectives where we attempt to model a system in the context of remote parameters. Therefore such remote effects may have a principal point of action removed from the point of interest, however their influence is still highly significant there.

1.2.1 Mass and inertia

Mass is an elementary property capable of acquiring or expending kinetic energy within the system due to imposed velocity gradients. Weight is a force quantity and is the effect of gravitational acceleration acting on mass. In the same way a mass will accelerate in the direction of (and under the influence of) a force, or resultant force in a force system. We derive the work quantity from the product of force and displacement in the direction of the force. Work on a mass is commensurate with the overall change in kinetic energy. Accordingly the concept of mass is central to Newton's laws of motion upon which the above statements are founded. We conventionally regard mass as inelastic, and add on our notions of elasticity and dissipation later as though they are physically separable from one another within the system. This simplistic view of things fades to some extent the nearer one gets to a continuous model, however it is a perfectly valid approach in lumped parameter representations.

Interpretation of effective mass in an engineering system depends on how we can break it down into discrete masses and stiffnesses. If several identifiable mass elements are clearly connected by spring-like elements, whose individual masses are negligible, then the problem is likely to be multi-degree of freedom in nature. Alternatively rigid interconnections between isolated lumped masses compel us to evaluate an effective mass quantity which can then be regarded as acting at any convenient point in the system. It is usual to refer the effective mass to a particular point of interest. Some well-known illustrations of how we use effective mass are, for example, natural frequency evaluation in a single degree of freedom mass and spring system where the spring mass is not negligible, or similar analysis of an oscillating motion conversion linkage as found in an engine valve gear configuration. Many such cases can be found in the literature (Tse, Morse and Hinkle, 1978; Rao, 1984) and involve expressing the system kinetic energy in a form equal to an equivalent kinetic energy at the chosen point of interest. This approach works equally well for translating and rotating systems (or combinations of these).

Discrete systems with many degrees of freedom naturally require a rather more complicated mathematical model than single degree of freedom problems, and the conventional approach here is to express the mass constituents of the system in the form of a matrix. In its simplest form the mass matrix does not depict any form of coordinate coupling, since in such cases all non-leading diagonal terms will be zero. More complicated problems will exhibit non-zero terms in other elements of the matrix; this points towards dynamic or inertia coupling between the coordinates. It is worth mentioning that identical considerations also apply to stiffness, and any coordinate coupling in this sense is called static or elastic coupling. An absence of both dynamic and static coupling indicates a situation where the resulting equations of motion can be solved independently. Coordinates in uncoupled problems are termed principal coordinates. Mass considerations in the domain of continuous systems introduce the idea of generalized mass, a quantity which is calculable for any particular mode of vibration. So, in a continuous system the mass which is operative in the system is a function of the mode shape and as such the resulting 'mass' will be seen to vary substantially between the modes.

1.2.2 Stiffness

All vibrating systems possess some form of elasticity, and in its most elementary form this is regarded as a simple linear proportionality relating displacement to applied load. Springs may assume a variety of physical forms and helical, torsional and flat springs are commonplace. The mass of a spring, or spring-like element, is often neglected in elementary analysis, however it can frequently be found to be contributory to a sizeable correction in natural frequency estimation.

Deformation of a spring arises because of an applied force so that there is a relative displacement between the extremities of the spring. The magnitudes of the force and displacement in a linear spring are related by a simple constant of proportionality sometimes known as the rate or stiffness. The units of stiffness are conventionally N/m. A deformed spring stores up potential or strain energy, this is equal to the work done in deformation. Stiffness is an additive quantity in the sense that several connected stiffnesses will have an overall combined effect. Simple rules apply, and a restatement and elaboration of these is unnecessary here; substantial coverage of simple spring equivalence problems is attributed to many authors, (Den Hartog, 1956; Timoshenko *et al.*, 1974; Tse *et al.*, 1978; Thomson, 1981; Rao, 1984; Meirovitch, 1986). Potential energy considerations also call for an appraisal of mass positioning, whereby an additional *mg* term can sometimes be added to the strain energy term.

Linearity in springs is only guaranteed for small deflections and after a specific point on the load deflection plane we find that the simple proportionality rule no longer applies. It is possible to linearize large deflection problems, however the errors introduced can be excessively high. Further discussion on the subject of nonlinear springs will be found in the second section of this book.

1.2.3 Damping

Damping is a very complicated and specialized subject in itself but one to which we must devote a certain amount of effort in understanding if we are to competently incorporate it in our analysis. Relatively few texts are available which deal with the subject in the depth it deserves, but the reader is referred to two authoritative books (Lazan, 1968; Nashif *et al.*, 1985), which are exceptions to this rule and which deal with the many different facets of structural and machine damping in a practical and useful manner. There is also much useful information on damping and its mechanisms in the excellent recent text on nonlinear oscillations (Nayfeh and Mook, 1979).

It is well known that lightly damped structures dissipate their energy more slowly than their heavier damped counterparts and therefore continue in motion for longer, with a correspondingly slower decay in the magnitude of motion. Damping as a phenomenon does not appear only in oscillatory problems, but in all instances in which mechanical energy is expended in inciting a system into motion. Nashif *et al.*(1985) mention the indisputable but perhaps not immediately obvious example of the 'efficient' golfball, where the highly elastic inner material of the ball is specially designed to instantaneously absorb a huge amount of energy on impact with the club and then to release it as quickly as possible so that almost all this energy is utilized in propulsion. Clearly this is an example of minimal damping.

Classroom theoretical damping models usually assume the form of the classical linear viscous damper where the damping forces generated are proportional to the velocity gradient across the device. This oil–piston–dashpot model is rather contrived and not indicative of 'real' engineering. The nearest practical instance of the simple dashpot damper is the automotive shock absorber, but in most cases additional non-viscous effects and supplementary stiffnesses are introduced which take us further away from the simple $C\dot{x}$ model (where C is the coefficient of linear viscous damping and $\dot{x}$ the velocity across the damper). The measurement method that relates to viscous damping with its exponential decay characteristics, is the logarithmic decrement, by which we relate the amplitudes of two cycles n cycles apart in the form of

$$\delta = \frac{1}{n} \ln\left(\frac{x_1}{x_2}\right). \tag{1.1}$$

It is a comparatively easy matter to evaluate the damping coefficient C, or alternatively the damping ratio ξ, where $\xi = C/C_c$, given that C_c is the critical damping constant, and we can show that

$$\delta = \frac{2\pi\xi}{\sqrt{1 - \xi^2}} = \frac{\pi C}{m\omega_d}, \tag{1.2}$$

where m and ω_d represent the mass and the damped natural frequency of free vibration. Viscous damping serves to limit resonant motion and is easily incorporated into most mathematical models.

Other definable linear damping varieties are dry friction, hysteretic and acoustic damping. Dry friction, or Coulomb damping, can be found in situations where relative motion between two contacting surfaces introduces a frictional force so that a single degree of freedom problem could be modelled by

$$m\ddot{x} + \mu N \operatorname{sgn} \dot{x} + kx = F(t), \tag{1.3}$$

where we assume $F(t)$ is a harmonically varying force and μ and N are the coefficient of friction and some prescribed static normal force respectively. If $F < \mu N$ then there will clearly be no motion of mass m, however when we reach $F > \mu N$ continuous oscillatory motion will occur (and the frictional force will change sign with $\dot{x}$). This type of damping does not necessarily always provide a limitation on resonant amplitude and the reader is referred to the books of Lazan (1968) and Nashif *et al.* (1985) for further insight.

Hysteretic damping is a form of material damping and this differs from the damping forms discussed previously since these are essentially the result of structural configuration. All elastic materials will give rise to a 'loading–unloading' loop in the force–displacement or stress–strain plane. This loop is called the hysteresis loop and the exact geometrical nature of the loop depends on the material being investigated. The important behavioural difference between viscous and hysteretic damping is that the former dissipates its cyclic energy linearly with frequency, whereas the latter is independent of frequency and the damping is seen as a complex stiffness constituent. So for complicated problems one can assume a complex modulus which contains both a stiffness part (real) and a damping part (imaginary), without having to differentiate clearly between them in the physical problem.

This form of damping may be found in elastomeric and polymeric compounds, and in a structural sense it can occur through impact in gapped joints and also in some cases of joint slippage.

The fundamental damping quantity for the hysteretic case is the loss factor η, so we write the complex stiffness–damping modulus as

$$k' = k(1 + i\eta) \tag{1.4}$$

and in the case of a single degree of freedom problem we can put

$$m\ddot{x} + k'x = F(t). \tag{1.5}$$

We can compare the frequency independence of hysteretic damping with the strongly frequency dependent behaviour of the viscous case, as modelled by the equation

$$m\ddot{x} + c\dot{x} + kx = F(t). \tag{1.6}$$

By integrating the two cyclic products of force and displacement we can therefore arrive at the cyclic energy dissipation due to the two systems

$$D_V = \pi C\Omega A_1^2, \tag{1.7}$$

$$D_H = \pi k\eta B_1^2, \tag{1.8}$$

where A_1 and B_1 are assumed particular solution amplitudes for the viscous and the hysteretic cases respectively, and D_V and D_H are the viscous and hysteretic cyclic energies. Figure 1.1 (after Nashif *et al.*, 1985) shows how the hysteretically damped response curves for a single

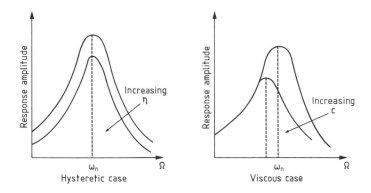

Figure 1.1 Response amplitude as a function of excitation frequency for hysteretic damping (after Nashif, Jones and Henderson, 1985) and for viscous damping.

degree of freedom problem all lie more or less perfectly symmetrically about the undamped natural frequency. Increasing η merely reduces the magnitude of the response at resonance. In the viscous case we obviously get a sizeable reduction in response amplitude, however there is also a marked tendency for the damped curves to centre themselves about frequencies which decrease from the undamped natural frequency as C increases. One can show that this dependency is governed by

$$\omega_d = \omega_n \sqrt{1 - 2\xi^2}, \tag{1.9}$$

where ω_n and ω_d are the undamped and damped natural frequencies respectively.

Acoustic radiation damping is an effect which will always occur to some extent when structural vibrations are generated in a surrounding fluid. This fluid can be gaseous or liquid, and the dissipation mechanism will predominantly be acoustical, with thermal effects perhaps contributing depending on the fluid viscosity.

Assessment of the physical damping effects is complicated by the combined influence of fluidic density, the velocity of sound within the fluid and also the structural mass and stiffness distributions. Although acoustic damping (sometimes called air damping) will always be present in vibrating structures it is rarely the main source of damping unless the fluid viscosity is high (not the case for air) or the structure is very lightweight. An exception to this is the case of very large amplitude responses for which damping can easily become nonlinear, and air damping may well at this point begin to predominate.

Another fluid damping phenomenon is pumping where the structure or machine is contained in some way by a nearly airtight enclosure and the pressures within cause a relatively large flow through the leak or orifice. This form of air damping is used in certain automotive shock absorber applications.

In conclusion it should be pointed out that damping (both linear and nonlinear) can be negative so that the system is self-exciting and will tend to contribute to its own response up to a point. In practice self-excitational systems will often contain an element of nonlinear negative damping which will aid vibratory motion or small amplitudes and then change over to a limiting effect as the amplitude begins to get larger. This Van der Pol oscillatory behaviour will give rise to a limit cycle where the final response magnitude is bounded and is also independent of the original initial conditions. We consider nonlinear damping in Chapter 3.

1.3 THE CONSERVATIVE SINGLE DEGREE OF FREEDOM MODEL

The most fundamental problem in vibrations is that of the linear single degree of freedom conservative mass–spring system as shown in Fig. 1.2. We assume one active coordinate and a massless spring and inelastic mass, and, because the problem in its basic form is conservative the governing equation of motion is simply

$$m\ddot{x} + kx = 0. \tag{1.10}$$

This is a result of applying Newton's second law and it is a linear, homogeneous, differential equation with constant coefficients. We can find an exact solution to this which describes harmonic, undamped, free vibration of the mass, and this will be of the form

$$x(t) = A\cos\omega t + B\sin\omega t. \tag{1.11}$$

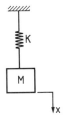

Figure 1.2 Conservative mass–spring system with one degree of freedom.

This is the general solution and it fully describes the resulting oscillatory motion of the problem. We can rewrite equation (1.11) as

$$x(t) = C \cos(\omega t - \phi), \qquad (1.12)$$

where the single constant C is the amplitude of response and is given by

$$C = (A^2 + B^2)^{1/2}, \qquad \text{and the phase angle } \phi \text{ is } \tan^{-1}(B/A). \quad (1.13)$$

Equation (1.12) gives us a response amplitude as well as a phase angle between response and a hypothetical excitation. We can evaluate A and B, and hence C, by referring to the initial conditions that apply. These are the displacement and velocity at time zero ($x(0)$ and $\dot{x}(0)$) which may be substituted into equation (1.11) and its first time-derivative to yield

$$A = x(0) \qquad \text{and} \qquad B = \frac{\dot{x}(0)}{\omega}. \qquad (1.14)$$

And so we have

$$x(t) = x(0) \cos \omega t + \frac{\dot{x}(0)}{\omega} \sin \omega t, \qquad (1.15)$$

which gives us a mathematical description of the ensuing response. Equation (1.15) is only valid at and after 'time zero' (whatever that may be). Any excitational effects that are acting on the system, and causing it to vibrate, immediately cease at that point. We are then left with the 'initial conditions'. In addition to this it is important to remember that an absence of dissipatory forces will mean that the resulting solution will continue oscillating at a constant amplitude for all time. Such are the limitations of this fundamental model.

1.3.1 Energy within the system

The conservative problem of equation (1.10) is defined as one which maintains a constant energy level. The energies within the system comprise the kinetic energy of the oscillating mass and the potential energy of the spring due to its deformation. We ascribe the variables T and U to these two energies and write

$$\frac{dE}{dt} = 0 \qquad \text{where} \qquad E = T + U, \qquad (1.16)$$

so that the total energy of the system, E, remains constant for all time and the system is conservative.

The kinetic energy of a mass m moving with velocity $\dot{x}$ is given by

$$T = \tfrac{1}{2} m \dot{x}^2, \qquad (1.17)$$

therefore substitution of equation (1.12) into (1.17) leads to

$$T_{max} = \tfrac{1}{2}m(-\omega C)^2, \tag{1.18}$$

where we can write T_{max} instead of T because C is the peak response amplitude. Also the potential energy will be

$$U = \tfrac{1}{2}kx^2 \tag{1.19}$$

and so equation (1.12) yields

$$U_{max} = \tfrac{1}{2}kC^2. \tag{1.20}$$

Kinetic energy will be at a maximum when the potential energy is at a minimum (i.e. zero) and vice versa. This provides a physical basis for the cyclic nature of harmonic motion and as a result of this one can say that T_{max} must equal U_{max} in a conservative system, and so we can equate equations (1.18) and (1.20) to give

$$m\omega^2 C^2 = kC^2. \tag{1.21}$$

Therefore the frequency of oscillation of the system ω is defined

$$\omega = \sqrt{\frac{k}{m}} \tag{1.22}$$

and this frequency is called the undamped natural frequency of free vibration. This result for ω is noteworthy and justifies our brief perusal of the conservative single degree of freedom problem because it forms an important general stage in our understanding. It can be extended to more complicated models such as the case of the multi-degree of freedom problem in which the known stiffness and mass matrices can be substituted into equation (1.22) in order to compute a natural frequency estimate.

1.4 THE NON-CONSERVATIVE SINGLE DEGREE OF FREEDOM MODEL

We can now quickly step up the level of modelling realism, and initially we do this by introducing non-conservative forces in the guise of damping and a harmonic exciting force. A forced-damped system such as this is shown in Fig. 1.3. Derivation of the equation of motion is achieved by considering the forces acting on the mass when it is regarded as a free body. The excitation is shown as acting downwards and since the mass is hanging from the fixture via the spring and damper it is clear that these forces, respectively proportional to displacement and velocity, will oppose the forced motion of the system. By adopting the convention of downwards action as positive and by taking Newton's second law, which postulates that the rate of change of momentum (or

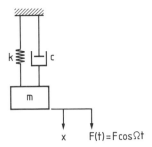

Figure 1.3 Forced-damped single degree of freedom system.

$m\ddot{x}$ for constant mass) is proportional to the vertical system forces, we get

$$m\ddot{x} = -kx - c\dot{x} + F\cos\Omega t. \qquad (1.23)$$

The effect of weight (mg) is exactly counteracted by a static spring extension, thus the static problem is subsumed into the dynamic one and need not concern us. Equation (1.23) is conventionally written with all but the excitation term on the left-hand side,

$$m\ddot{x} + c\dot{x} + kx = F\cos\Omega t. \qquad (1.24)$$

The solution to this equation is in two additive parts; the complementary function and the particular integral. The complementary function can be investigated independently of the particular integral, and it describes the decaying transient motion of the system for which $F(t) = 0$. Since the complementary function is decaying, and system dissipation here is assumed to be due to the viscous damping c, this part of the general solution can tell us much about the actual level of damping active within the system.

1.4.1 Complementary functions

Therefore we take

$$F(t) = F\cos\Omega t = 0 \qquad (1.25)$$

and assume that x will change exponentially with time so that we have

$$x = P\,e^{st}, \qquad (1.26)$$

where P and s are constants. Differentiation of (1.26) as necessary, and substitution of (1.25) into equation (1.24) yields an algebraic equation of the form

$$P\,e^{st}(ms^2 + cs + k) = 0 \qquad (1.27)$$

for which $P e^{st}$ will not always be zero, thus we must state that

$$ms^2 + cs + k = 0. \tag{1.28}$$

This is a straightforward quadratic equation in s for which there will be two roots

$$s_{1,2} = -\frac{c}{2m} \pm \sqrt{\frac{c^2}{4m^2} - \frac{k}{m}}. \tag{1.29}$$

The existence of two roots means that we must extend solution (1.26) to

$$x = P_1 e^{s_1 t} + P_2 e^{s_2 t} \tag{1.30}$$

and we will need to use our initial conditions to find P_1 and P_2. The notion of critical damping (briefly alluded to in Section 1.2.3 in reference to equation (1.2)) is usefully employed at this point and may be succinctly defined as that damping which modifies the free vibration response from periodic to non-periodic in form. This transitional case occurs when the terms under the radical of (1.29) equal zero. Therefore

$$\frac{c}{4m^2} - \frac{k}{m} = 0; \qquad c = c_c. \tag{1.31}$$

From this we get

$$c_c = 2m\left(\frac{k}{m}\right)^{1/2} = 2m\omega, \tag{1.32}$$

where $\omega = \omega_n$; the suffix n is added to emphasize that this is the natural frequency of free undamped vibration.

As $\xi = c/c_c$ one can write, using (1.32),

$$c = 2m\omega_n\xi. \tag{1.33}$$

Equation (1.29) can be stated more concisely by incorporating (1.22) and (1.33) to give

$$s_{1,2} = (-\xi \pm \sqrt{\xi^2 - 1})\omega_n. \tag{1.34}$$

At this stage we can substitute for $s_{1,2}$ in equation (1.30) to arrive at a solution of the following form

$$x = P_1 \exp(-\xi + \sqrt{\xi^2 - 1})\omega_n t + P_2 \exp(-\xi - \sqrt{\xi^2 - 1})\omega_n t. \tag{1.35}$$

This is of little use as it stands because P_1 and P_2 are still not explicitly related to measurable system parameters.

Progress can be made here by considering three 'broad' damping cases; these are the underdamped, critically damped and overdamped conditions. Taking the underdamped case first we stipulate that this is defined by

$$c < c_c, \qquad \therefore \ \xi < 1. \tag{1.36}$$

If this is the situation then the radical of (1.34) will be negative, consequently roots (1.34) will be written as

$$s_1 = (-\xi + i\sqrt{1 - \xi^2})\omega_n \quad \text{and} \quad s_2(-\xi - i\sqrt{1 - \xi^2})\omega_n. \quad (1.37)$$

The roots s_1 and s_2 are complex and may be substituted into equation (1.30) such that with some further manipulation we get

$$x = (P_1 \exp(i\sqrt{1 - \xi^2}\ \omega_n t) + P_2 \exp(-i\sqrt{1 - \xi^2}\ \omega_n t)) \exp(-\xi\omega_n t). \quad (1.38)$$

We now have to consider the initial conditions for the system in order to determine P_1 and P_2.

Equation (1.38) is restated in the following form

$$x = [(P_1 + P_2)\cos\sqrt{1 - \xi^2}\omega_n t$$
$$+ i(P_1 - P_2)\sin\sqrt{1 - \xi^2}\ \omega_n t]\exp(-\xi\omega_n t) \quad (1.39)$$

and we can put

$$Q_1 = P_1 + P_2,$$
$$Q_2 = i(P_1 - P_2), \quad (1.40)$$

which gives

$$x = [Q_1 \cos\sqrt{1 - \xi^2}\ \omega_n t + Q_1 \sin\sqrt{1 - \xi^2}\ \omega_n t]\exp(-\xi\omega_n t). \quad (1.41)$$

At $t = 0$ we can state, therefore, that

$$x(0) = x_0; \quad \dot{x}(0) = \dot{x}_0 \quad (1.42)$$

and substitution of these conditions into equation (1.41) yields

$$Q_1 = x_0 \quad \text{and} \quad Q_2 = \frac{\dot{x}_0 + \xi\omega_n x_0}{\sqrt{1 - \xi^2}\ \omega_n}. \quad (1.43)$$

The results for Q_1 and Q_2 may now be substituted into equation (1.41) to reveal the full result for the complementary function part of the general solution for the underdamped case. It is perhaps more convenient to express the solution (1.41) in the following manner

$$x = X \exp(-\xi\omega_n t)\cdot\cos(\omega_d t + \phi), \quad (1.44)$$

where

$$X = (Q_1^2 + Q_2^2)^{1/2} \quad \text{and} \quad \phi = \tan^{-1}\left(\frac{Q_2}{Q_1}\right) \quad (1.45)$$

and ω_d is as in equation (1.9).

The complementary function (c.f.) part of the solution describes motion which is periodic (with frequency ω_d) but which decays exponentially due to $\exp(-\xi\omega_n t)$. This rate of decay is the clue to the active damping, and the logarithmic decrement method (equations (1.1) and (1.2)) may be used to evaluate ξ. The oscillatory response of an underdamped system is shown in Fig. 1.4(a).

The critically damped case requires that

$$c_c = c, \quad \text{and so} \quad \xi = 1$$

thus equation (1.34) reduces to

$$s_{1,2} = -\omega_n. \tag{1.46}$$

Since $\xi \to 1$ we have $\omega_d \to 0$ and so equation (1.41) becomes

$$x = [Q_1 + Q_2\omega_d t] \exp(-\omega_n t). \tag{1.47}$$

The explicit reference to ω_d is avoided (since it is tending to zero in the limit) by putting

$$R = Q_2\omega_d. \tag{1.48}$$

The initial conditions of (1.42) are used again to yield

$$Q_1 = x_0 \quad \text{and} \quad R = \dot{x}_0 + \omega_n x_0. \tag{1.49}$$

Thus the critically damped solution is

$$x = [x_0 + (\dot{x}_0 + \omega_n x_0)t] \exp(-\omega_n t). \tag{1.50}$$

The response of a critically damped system is aperiodic and is depicted in Fig. 1.4(b).

The final case of overdamping is treated by letting

$$c > c_c, \quad \text{and so} \quad \xi > 1,$$

which results in the roots (1.34) being real and distinct.

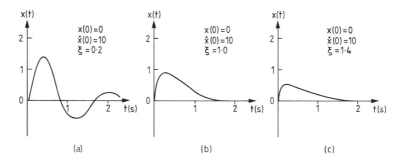

(a)   (b)   (c)

Figure 1.4 Damped, free vibration responses of a single degree of freedom system: (a) underdamped; (b) critically damped; (c) over-damped.

From this we have

$$x = P_1 \exp(-\xi + \sqrt{\xi^2 - 1})\omega_n t + P_2 \exp(-\xi - \sqrt{\xi^2 - 1})\omega_n t. \quad (1.51)$$

Identical initial conditions are used to determine the constants P_1 and P_2 such that

$$P_1 = \frac{x_0 \omega_n (\xi + \sqrt{\xi^2 - 1}) + \dot{x}_0}{2\omega_n \sqrt{\xi^2 - 1}}, \quad (1.52)$$

$$P_2 = \frac{-x_0 \omega_n (\xi - \sqrt{\xi^2 - 1}) - \dot{x}_0}{2\omega_n \sqrt{\xi^2 - 1}}. \quad (1.53)$$

Again we will have aperiodic motion which decays exponentially with time, as shown in Fig. 1.4(c).

1.4.2 The particular integral

We return to equation (1.24) in order to derive the particular integral (P.I.) part of the solution, and as a first step we take a generalized particular solution which will be a harmonic, and perhaps of the following form

$$x = X_0 \cos(\Omega t - \phi_0). \quad (1.54)$$

Substitution of this in equation (1.24) generates two equations in which we can solve for X_0 and ϕ_0, thus,

$$X_0[(k - m\Omega^2) \cos \phi_0 + c\Omega \sin \phi_0] = F, \quad (1.55)$$

$$X_0[(k - m\Omega^2) \sin \phi_0 - c\Omega \cos \phi_0] = 0, \quad (1.56)$$

leading to

$$X_0 = \frac{F}{[(k - m\Omega^2)^2 + c^2\Omega^2]^{1/2}} \quad (1.57)$$

and

$$\phi_0 = \tan^{-1}\left[\frac{c\Omega}{k - m\Omega^2}\right]. \quad (1.58)$$

The full response is given by c.f. + P.I., and so we combine equations (1.44), (1.54) and (1.57) to give

$$X_F = X \exp(-\xi\omega_n t) \cdot \cos(\omega_d t + \phi) + X_0 \cos(\Omega t - \phi_0), \quad (1.59)$$

where X_F is the 'full solution'.

To conclude, the non-conservative single degree of freedom problem can be fragmented into two parts; the complementary function which describes the decay of the free transient (case where $F(t) = 0$) and the particular integral which caters for that part of the solution governed by a harmonic excitation. The roots of the characteristic equation (1.28) assume different forms depending on the degree of damping acting in the system and it is possible to represent and examine these graphically as well as analytically as in the summarized cases given above. The root-locus diagram is used for this, and depicts the enumeration of ξ in the complex plane. The locus of the roots of the characteristic equation is a semicircle symmetrical about the horizontal real axis; Fig. 1.5 shows how we can relate the roots $s_{1,2}$ to the damping which is operative.

Critical damping is of most practical significance as it represents the minimum level of damping necessary for aperiodic motion, and therefore combines a necessity for energy dissipation with (theoretically) zero overshoot of the mass in motion. Some typical applications of critical damping are in gun-carriage and moving coil instrument design.

The introduction of a harmonic forcing term involves the particular integral part of the solution, for which equations (1.54), (1.57) and (1.58) apply. The condition of resonance arises when $\Omega = \omega_n$, and so equation (1.57) reduces to

$$X_0 = \frac{F}{c\omega_n} \tag{1.60}$$

and ϕ_0 (equation (1.58)) equals $\pi/2$ radians exactly. The resonant condition is where the response amplitude reaches a maximum; this will tend to approach infinity as c tends to zero. Conversely we find that the response amplitude decreases significantly with increasing c. Specimen amplitude–frequency and phase angle–frequency relationships are given in Figs 1.6 and 1.7.

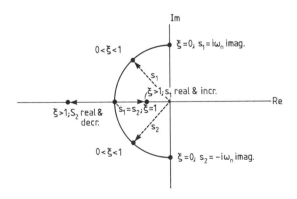

Figure 1.5 Locus of the roots of characteristic equation (1.28).

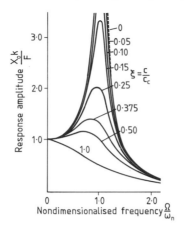

Figure 1.6 Forced-damped single degree of freedom system responses, showing the effect of varying damping ratio ξ.

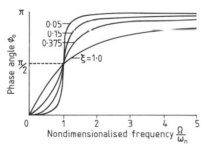

Figure 1.7 Phase angle as a function of non-dimensionalized excitation frequency ratio showing the effect of varying damping ratio ξ.

1.5 PRACTICAL USES FOR SINGLE DEGREE OF FREEDOM MODELS

We can use these fundamental representations for a variety of practical modelling and design exercises. In particular one can frequently obtain helpful initial information on structural and machine natural frequencies, vibrational transmissibility, effectiveness of simple dampers and absorbers, simplified vehicle body interactions with their suspensions, component matching of rotating plant, scientific instrument design and rotational unbalance problems.

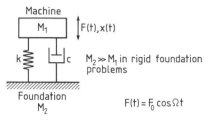

Figure 1.8 Vibration isolation schematic.

1.5.1 Isolation problems

A single degree of freedom model is adequate in the case of the plant–foundation interaction problem where the objective is to eliminate, or at least minimize, the transmission of plant generated vibration through to the factory floor. The equipment (for example a high speed punch press) is mounted on damped spring mounts so that dynamically one obtains the system of Fig. 1.8. The proviso here is that the mass of the floor, M_2, is very much greater than that of the machine, M_1, so that the problem does not tend to one of two degrees of freedom. The harmonic forcing function is a result of the 'flywheel effect' and the 'power stroke', and we assume that the machine runs at one fixed speed. Clearly the equation of motion will be identical to equation (1.24) and the particular solution will be as in equation (1.54). The forces transmitted through the spring and damper parts of the isolator will be, respectively,

$$F_s = kx = kX_0 \cos(\Omega t - \phi_0), \tag{1.61}$$

$$F_d = c\dot{x} = -c\Omega X_0 \sin(\Omega t - \phi_0). \tag{1.62}$$

Consideration of these forces vectorially gives the total transmitted force magnitude as

$$F_t = \sqrt{F_s^2 + F_d^2} = X_0 \sqrt{k^2 + (c\Omega)^2}. \tag{1.63}$$

Transmissibility is defined as the ratio F_t/F and is denoted by T_R, therefore by combining equations (1.57) and (1.63) we obtain

$$T_R = \sqrt{\frac{k^2 + (c\Omega)^2}{(k - m\Omega^2) + (c\Omega)^2}}. \tag{1.64}$$

This can be non-dimensionalized to give

$$T_R = \sqrt{\frac{1 + [2\xi(\Omega/\omega_n)]^2}{[1 - (\Omega/\omega_n)^2]^2 + [2\xi(\Omega/\omega_n)]^2}}. \tag{1.65}$$

The effects of variable frequency and damping on transmissibility are shown in Fig. 1.9.

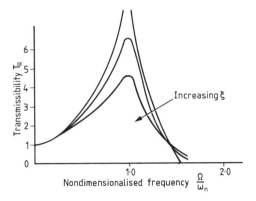

Figure 1.9 The effects of variable excitation frequency and damping on the transmissibility of vibration.

The problem is extended somewhat if the foundation mass m_2 is significantly reduced and incited into motion by means of coupling with m_1. The problem is then one of two degrees of freedom.

1.5.2 Vehicle problems

If one constrains the vehicle body problem of Fig. 1.10(a) to vertical motion such that there is one independent coordinate it can easily be reduced to the equivalent form of Fig. 1.10(b) and a limited, vertical ride characteristic deduced. Clearly though, any rotational motion could not be accounted for in the model and in reality one might prefer, say, a three degree of freedom model where pure vertical translation is coupled to transverse and longitudinal rotations about some point, not

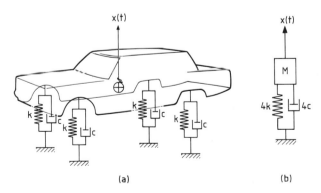

Figure 1.10 Vehicle body problem: (a) vertical ride with no pitch or roll; (b) equivalent lumped form.

necessarily the centre of mass. This extended model would involve static and dynamic coupling of three coordinates. The usefulness of the single degree of freedom approach here is, therefore, rather circumscribed.

We can also use our simple single degree of freedom theory to reduce the vibrations transmitted to the vehicle body by the engine as a result of its almost-periodic torque delivery characteristics, and to this end we mount the engine on rubberized blocks which serve to isolate the vehicle from torsional oscillations. There will also be vertical and horizontal components of inertia force which will induce translations at the mounting points and one would have to extend the number of degrees of freedom of the model to cater for these. For the case of torque induced vibration we would need to ensure that the equivalent stiffness of the mountings provided a natural frequency of free vibration of the engine in rotation about the axis of torque that is very much lower than $nN/180$ Hz, where n is the number of cylinders and N is the running speed in rev/min. In this way the transmissibility of the system can be kept at a low level. Torque fluctuations are also of concern at the engine design stage and an additional inertia in the form of a flywheel together with a tuned absorber are methods which can be used to smooth out power delivery. The well-known Sarazin–Chilton pendulum absorber is an illustration of the latter case and is really just an ingeniously designed centrifugal pendulum with a high pendulum mass and a small pendulum length, and it is this that makes it so applicable to direct mounting on the crankshaft 'flywheel' counterbalance. The Sarazin–Chilton absorber of Fig. 1.11 is little more than a single degree of freedom pendulum problem under the influence of a centrifugal force field, and works by offsetting the disturbing torque by the creation of an equal and opposite pendulum inertia torque.

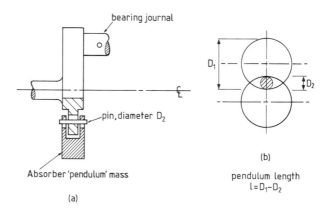

Figure 1.11 The Sarazin–Chilton absorber (after Den Hartog, 1947).

1.5.3 Design of vibration detection instruments

The piezo-electric accelerometer contains a mass and a piezo-crystal which develops a polarized charge proportional to the imposed mechanical strain. By incorporating a viscous fluid around the mass one can regard the device as a single degree of freedom system incorporating stiffness (the crystal) and damping; this is shown in Fig. 1.12. Accelerometers are designed to have a flat response over the working frequency range and so the natural frequency of the device should be well above this. One of the principal advantages of the accelerometer over other devices is that almost no phase distortion of the measured signal occurs in practice, which means that the output from the accelerometer is virtually identical to the quantity being measured. A recent innovation in accelerometer design has been the piezo-plastic accelerometer where the crystalline transducing element is replaced by a low cost plasticized version. These accelerometers have a direct output and do not need the usual expensive external charge amplifier.

The accelerometer as a single degree of freedom instrument is described as seismic; this merely means that a single lumped and sprung mass is used to detect vibrations, and similarly on a larger scale we have the seismologist's seismograph in which the suspended mass is frequently over 1000 kg giving a fundamental <0.1 Hz, dependent on the suspension stiffness.

1.5.4 Applications to rotational plant

Rotating systems in general can undergo either torsional oscillatory motion, where a small torsional oscillation is superimposed on the rotational speed, or lateral **whirling**. The first case occurs in situations where there is a cyclic, or non-steady, torque delivery from the prime mover (as discussed in the latter half of Section 1.5.2) so that we have

$$\theta = \Theta \sin n\Omega t \qquad (1.66)$$

superimposed on a rotational speed Ω, where n may be unity or an integer depending on the torque fluctuation characteristic. Torque

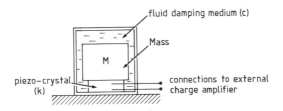

Figure 1.12 Piezo-electric accelerometer schematic.

fluctuations are an unavoidable feature of reciprocating engine driven plant installations and although primary and secondary balancing of the engine will reduce them it will, on occasion, be necessary to incorporate a damper such as the Sarazin–Chilton device or the crankshaft-end mounted fluid-filled Houde device. Reduction of Θ may also be achieved simply by the addition of more inertia, as in the case of a coupled generator or pumping system for example.

Lateral vibrations or whirling, tend to occur in specific locations within a plant installation and are in many cases eminently suitable for single degree of freedom analysis in an effort to 'design out' their potentially destructive effects. As a rule we do not consider the inertial properties of the shaft itself but take the rotor as the inertial element (in whatever form this may assume) and assign a lateral stiffness to the shaft. A single degree of freedom model on this basis will provide a reasonable estimate of the fundamental natural frequency or critical speed in certain cases, particularly when the lumped rotor inertia is significantly higher than the distributed inertia of the shaft. Whirling only happens when the mass centre of the system is not coincident with the geometric centre and we assume that a degree of unbalance in the rotor (a heavy gear or sprocket on a flexible shaft, or a turbine rotor, or a motor armature for example) will create such a condition. Figure 1.13 shows this in exaggerated form. Vertical and horizontal resolution of forces produces two equations of motion (in x and y)

$$m \frac{d^2}{dt^2} (x + a \cos \Omega t) + c\dot{x} + kx = 0, \qquad (1.67)$$

$$m \frac{d^2}{dt^2} (y + a \sin \Omega t) + c\dot{y} + ky = 0, \qquad (1.68)$$

assuming linear velocity dependent damping c.

Since whirl motion is circular about the rotation centre A the x and y response amplitudes will be equal in magnitude and equal to the radius of the described circle itself.

Solutions of the form of (1.54) are taken and we get

$$r = X_0 = Y_0 = \frac{ma\Omega^2}{k\sqrt{(1 - (\Omega/\omega_n)^2)^2 + (2\xi(\Omega/\omega_n))^2}} \qquad (1.69)$$

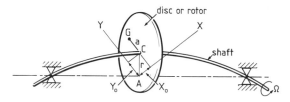

Figure 1.13 Out of balance rotor.

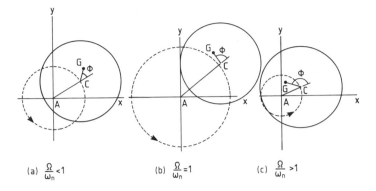

Figure 1.14 Whirl orbits of out of balance rotor, three cases: (a) $(\Omega/\omega_n) < 1$; (b) $(\Omega/\omega_n) = 1$; (c) $(\Omega/\omega_n) > 1$.

after substitution into equations (1.67) and (1.68), where r = whirl circle radius, ω_n = natural frequency of lateral shaft vibration, and $\xi = c/2\omega_n$.

The physical significance of the frequency ratio Ω/ω_n is distinctly shown in Fig. 1.14 where the three cases of $\Omega/\omega_n < 1$ (but still close), $\Omega/\omega_n = 1$ and $\Omega/\omega_n > 1$ are given.

All rotating installations will generate critical whirlspeed behaviour to some degree and one should take the utmost care to prevent whirlspeed coinciding with running speed or any dominant harmonics that may be present. One way of reducing the critical speed is to install flexible support bearings for the assembly. In the case outlined the fatigue problem is perhaps not as serious as in certain multi-whirl speed cases because there will be no stress reversals, and so the compressive and tensile sides of the bowed out shaft will remain in those respective states. However if the operating speed is between two reasonably close critical speeds it is likely that there will be a severe fatigue problem due to the generation of two stress reversals per shaft revolution.

1.6 TRANSIENTS AND IMPULSES IN SINGLE DEGREE OF FREEDOM SYSTEMS

Harmonic forcing functions can often closely simulate real situations, but there are also instances when the machine or structure can be excited into a vibratory response by some suddenly applied impulse, or a non-periodic waveform, and in these cases the resulting non-steady state response is called a transient response. The transient response always dies away, with the decaying oscillation occurring at the system's natural frequency. There may also be a lasting and constant aspect to the response, but as a rule this will be identical to the excitation in form and will only be apparent because of high amplitude forcing. The conventional way of understanding transient behaviour is first of all to treat the impulse case, since a consecutive series of short impulses can lead to a reasonably clear representation of a non-periodic function. An alternative approach is through Fourier analysis. It must be said at this point that both these techniques require a fairly detailed mathematical treatment and so in the space available here an abbreviated development of the theory is presented.

1.6.1 The unit impulse

We first consider a rectangular impulse with an area equal to 1 and a width of T. By compressing the pulse width down to zero and maintaining the area at unity we establish the so-called **unit impulse**. The impulse amplitude will clearly be infinitely large. We can ensure this mathematically be defining the rectangular pulse height as $1/T$.

One defines the impulse, which occurs at time $t = 0$, by stating,

$$\delta(t) = 0 \qquad \text{for all} \quad t \neq 0. \tag{1.70}$$

Therefore,

$$\int_{-\infty}^{\infty} \delta(t)\,\mathrm{d}t = 1, \tag{1.71}$$

where δ is the impulse itself, or the Dirac delta function.

On the other hand the impulse could occur at a time $t = 0$ such as $t = \tau$ for which case the impulse function is

$$\delta(t - \tau) = 0 \qquad \text{for all} \quad t \neq \tau. \tag{1.72}$$

This is nothing more than a shift of the unit impulse through to τ along the time axis. From this we get

$$\int_{-\infty}^{\infty} \delta(t - \tau)\,\mathrm{d}t = 1. \tag{1.73}$$

Practical generation of a unit impulse is highly unlikely since it is defined as having zero width, unit area and infinite amplitude, however the concept can be an extremely useful approximation and we can consider very short duration pulses (such that $\varepsilon \ll 1/\omega_n$, where ε and ω_n are the pulse width and system natural frequency respectively) in the light of unit impulses. To this end we investigate the damped single degree of freedom oscillator with an impulsive excitation such that we have

$$m\ddot{x} + c\dot{x} + kx = \delta(t). \tag{1.74}$$

We assume that the initial conditions are zero ($\dot{x}(0) = x(0) = 0$) and that the pulse length ε is very small (i.e. $\varepsilon \to 0$). Integration of equation

(1.74) with respect to time in the limit $\varepsilon \to 0$ gives

$$\lim_{\varepsilon \to 0} \int_0^\varepsilon (m\ddot{x} + c\dot{x} + kx)\,dt = \lim_{\varepsilon \to 0} \int_0^\varepsilon \delta(t)\,dt = 1. \qquad (1.75)$$

The left-hand side inertia term becomes

$$\lim_{\varepsilon \to 0} \int_0^\varepsilon m\ddot{x}\,dt = \lim_{\varepsilon \to 0} m[\dot{x}(\varepsilon) - \dot{x}(0)]. \qquad (1.76)$$

The damping term is

$$\lim_{\varepsilon \to 0} \int_0^\varepsilon c\dot{x}\,dt = \lim_{\varepsilon \to 0} c[x(\varepsilon) - x(0)] \qquad (1.77)$$

and the stiffness term is time-invariant and therefore

$$\lim_{\varepsilon \to 0} \int_0^\varepsilon kx\,dt = 0. \qquad (1.78)$$

Equation (1.76) can be further simplified by equating it to $m\dot{x}(0+)$ which represents the velocity change after the interval ε. Equations (1.77) and (1.78) both reduce to zero since there can be no actual displacement change in an interval of length $\varepsilon \to 0$. The problem is now essentially one of free vibration with damping but with revised initial conditions. Combining equation (1.75) and (1.76) leads to

$$\lim_{\varepsilon \to 0} m[\dot{x}(\varepsilon) - \dot{x}(0)] = 1. \qquad (1.79)$$

Thus

$$\dot{x}(0+) = \frac{1}{m}. \qquad (1.80)$$

Since $x(0)$ is unchanged, and equal to zero, on referring to equation (1.43) we find that

$$Q_1 = 0 \quad \text{and} \quad Q_2 = \frac{1}{m\sqrt{1 - \xi^2}\,\omega_n}. \qquad (1.81)$$

Substitution of equation (1.81) into equation (1.41) gives

$$x = \frac{1}{m\omega_n\sqrt{1 - \xi^2}} \exp(-\xi\omega_n t).\sin\omega_d t. \qquad (1.82)$$

We can write this as the impulsive response $h(t)$, $(h(t) \equiv x)$, so,

$$h(t) = \frac{1}{m\omega_d} \exp(-\xi\omega_n t).\sin\omega_d t \qquad (1.83)$$

(for all $t > 0$).
Correspondingly for $t < 0$, $h(t)$ will be zero.

One can construct a similar analysis based on the time-shifted function (1.73) such that we get

$$h(t - \tau) = \frac{1}{m\omega_d} \exp\left(-\xi\omega_n[t - \tau]\right) \cdot \sin \omega_d(t - \tau) \qquad (1.84)$$

(for all $t > \tau$).

1.6.2 The convolution integral

We are now at a point where we can incorporate the unit impulse theory into a useful representative model for some, as yet, arbitrary non-periodic excitation function $F(t)$. In Fig. 1.15 an $F(t)$ is shown as a function of time along with an impulse which lasts $\Delta\tau$. The overall response will be the product of the impulse magnitude (i.e. pulse area $F(\tau)\Delta\tau$) and the unit impulse response giving

$$\Delta x(t, \tau) = F(\tau)\Delta\tau \cdot h(t - \tau). \qquad (1.85)$$

The sum of the impulse responses to $F(\tau)$ will simply be

$$x(t, \tau) = \sum F(\tau)h(t - \tau)\Delta\tau. \qquad (1.86)$$

If we compress $\Delta\tau \to 0$ then the summation tends to an integral and we get

$$x(t, \tau) = \int_0^t F(\tau)h(t - \tau)\,d\tau \qquad (1.87)$$

into which one can substitute for $h(t - \tau)$ from equation (1.84).

The expression in (1.87) is known as the Convolution Integral.

We can use the convolution integral to find the response of undamped and damped systems to a step input, this being perhaps a long duration pulse (i.e. much longer than the period of the system).

1.6.3 Single degree of freedom response to a step input

In the absence of damping the undamped impulse response to the step

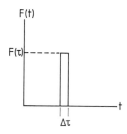

Figure 1.15 Impulse lasting $\Delta\tau$ for an arbitrary excitation function $F(t)$.

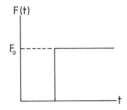

Figure 1.16 Step input function.

input of Fig. 1.16 becomes

$$h(t) = \frac{1}{m\omega_n} \sin \omega_n t. \tag{1.88}$$

If we substitute for $h(t - \tau)$ (which we easily obtain by time shifting through τ), in equation (1.87) we obtain

$$x(t, \tau) = \frac{F_0}{m\omega_n} \int_0^t \sin \omega_n(t - \tau) \, d\tau. \tag{1.89}$$

The integrand $\sin \omega_n(t - \tau)$ is simply a standard form and so we get

$$x = \frac{F_0}{m\omega_n{}^2} (1 - \cos \omega_n t). \tag{1.90}$$

This is the required solution to the undamped problem, and in a similar fashion we could use the convolution integral approach for the damped problem. The convolution integral is a powerful tool, and, depending on the nature of the forcing function and perhaps also if this is in conjunction with damping, it may sometimes be easier to numerically integrate the expression to get the answer. The response according to equation (1.90) to the unit step input ($F_0 = 1$) is given in Fig. 1.17 where it can be seen that the peak response to a unit step input is exactly twice the static deflection due to an input of $1/k$.

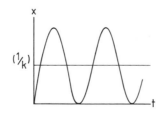

Figure 1.17 Oscillatory, undamped, single degree of freedom response to a step input function.

1.6.4 Fourier analysis

The Fourier series can be used to represent both periodic and non-periodic functions as it relies on an analysis of the harmonic components which constitute the function. To be more specific a convergent series of harmonic functions, where the individual function frequencies are integer multiples of some fundamental, may be found to closely model the non-sinusoidal function of interest.

The Fourier series is conventionally expressed as

$$F(t) = \frac{a_0}{2} + \sum_{n=1}^{\infty} (a_n \cos n\omega t + b_n \sin n\omega t), \qquad (1.91)$$

where the fundamental frequency is $\omega = 2\pi/T$ (and T is the period or the time it takes the function to get to the same point in the next 'cycle'), n is some positive integer and a_n and b_n are integrable functions of the excitation $F(t)$; these must exist and must be calculable. The Fourier coefficients are defined as

$$a_0 = \frac{2}{T} \int_0^T F(t)\,dt, \qquad (1.92)$$

$$a_n = \frac{2}{T} \int_0^T F(t) \cos n\omega t\,dt, \qquad (1.93)$$

$$b_n = \frac{2}{T} \int_0^T F(t) \sin n\omega t\,dt. \qquad (1.94)$$

An example in using the Fourier series could be one where we find the mathematical description for a square wave (which is obviously periodic but non-harmonic). The function is shown in Fig. 1.18 from which it can be seen that the peak amplitude is unity and the period T. For each repetitive cycle $F(t)$ is defined by

$$F(t) = 1, \qquad 0 < t < T/2,$$
$$F(t) = -1, \qquad T/2 < t < T. \qquad (1.95)$$

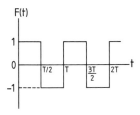

Figure 1.18 Square wave function with period T and unity peak amplitude.

This allows us to evaluate the Fourier coefficients leading to

$$a_0 = \frac{2}{T} \left[\int_0^{T/2} dt - \int_{T/2}^{T} dt \right] = 0, \tag{1.96}$$

$$a_n = \frac{2}{T} \left[\int_0^{T/2} \cos n\omega t \, dt - \int_{T/2}^{T} \cos n\omega t \, dt \right] = 0, \tag{1.97}$$

$$b_n = -\frac{1}{n\pi} \left\{ \left[\cos \frac{2n\pi}{T} t \right]_0^{T/2} - \left[\cos \frac{2n\pi}{T} t \right]_{T/2}^{T} \right\}. \tag{1.98}$$

Therefore,

$$b_n = \frac{4}{n\pi} \qquad \text{for odd } n,$$

and

$$b_n = 0 \qquad \text{for even } n. \tag{1.99}$$

Substitution of the Fourier coefficients back into equation (1.91) gives

$$F(t) = \frac{4}{\pi} \sum_{n=1,3,5,\dots}^{\infty} \sin \frac{2\pi nt}{T}. \tag{1.100}$$

The construction of the Fourier representation for the first four harmonics is as shown in Fig. 1.19.

For the non-periodic case it is convenient to rewrite the Fourier series in its complex form

$$F(t) = \sum_{n=-\infty}^{\infty} C_n e^{in\omega t} \tag{1.101}$$

in which the complex coefficients C_n are

$$C_n = \frac{1}{T} \int_{-T/2}^{T/2} F(t) e^{-in\omega t} \, dt, \qquad \text{for } n = 0, \pm 1, \pm 2, \dots \tag{1.102}$$

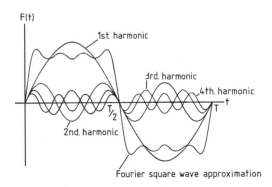

Fourier square wave approximation

Figure 1.19 Fourier approximation to a square wave function; the first four harmonics are summed according to equation (1.91).

and are obtained by writing $\cos n\omega t$ and $\sin n\omega t$ in their exponential forms thus

$$\cos n\omega t = \tfrac{1}{2}(e^{in\omega t} + e^{-in\omega t}), \tag{1.103}$$

$$\sin n\omega t = \frac{1}{2i}(e^{in\omega t} - e^{-in\omega t}). \tag{1.104}$$

The fundamental frequency ω is defined in equation (1.91). In order to 'stretch' the period, and in so doing effectively eliminate the periodicity we restate the series and its coefficients in the following form

$$F(t) = \frac{1}{2\pi} \sum_{n=-\infty}^{\infty} (TC_n) e^{i\omega_n t} \Delta\omega_n, \tag{1.105}$$

$$(TC_n) = \int_{-T/2}^{T/2} F(t) e^{-i\omega_n t}\, dt. \tag{1.106}$$

This is achieved by putting

$$n\omega = \omega_n , \tag{1.107}$$

and

$$\omega = \Delta\omega_n. \tag{1.108}$$

If T tends to infinity and we transform the discrete ω_n into the continuous ω equation (1.106) becomes, in the limit,

$$\underset{\substack{T\to\infty \\ \Delta\omega_n\to 0}}{\text{Lim}}\ (TC_n) = F(\omega) = \int_{-\infty}^{\infty} F(t) e^{-i\omega t}\, dt, \tag{1.109}$$

which enables us to rewrite equation (1.105) as

$$F(t) = \underset{\substack{T\to\infty \\ \Delta\omega_n\to\infty}}{\text{Lim}}\ \frac{1}{2\pi} \sum_{n=-\infty}^{\infty} (TC_n) e^{i\omega_n t} \Delta\omega_n. \tag{1.110}$$

Thus

$$F(t) = \frac{1}{2\pi} \int_{-\infty}^{\infty} F(\omega) e^{i\omega t}\, d\omega. \tag{1.111}$$

The integrals $F(t)$ and $F(\omega)$ are called the Fourier transform pair and together can be used to define some arbitrary non-periodic function. The proviso for the successful use of the Fourier transform is that the function $F(t)$ has no infinite discontinuities and has a finite number of minima or maxima in the domain $-\infty < t < \infty$. This is known as Dirichlet's condition.

In many cases the Fourier transform method will necessitate the evaluation of complicated and unwieldy integrals and for this reason is not always the best choice. Possibly its greatest strengths lie in finding a solution to non-deterministic problems where the forcing functions are random (or stochastic) and it is then more convenient to work in the frequency rather than the time domain. A very useful derivative method which applies to all forms of deterministic excitation function is the Laplace transform approach in which the decidedly tricky complex-plane integrations can be avoided. This method is recommended for practical engineering problems, as it enables a straightforward algebraic relationship to be established between the system excitation and response.

1.6.5 Laplace transforms and their application

The fundamental definition of the Laplace transform of a function $x(t)$ is

$$Lx(t) = \bar{x}(s) = \int_0^\infty e^{-st} x(t) \, dt, \tag{1.112}$$

where $Lx(t)$ and $x(s)$ are notation used to denote the transformed function. The variable s is complex and sometimes called the subsidiary variable, and e^{-st} is the transformation kernel. In brief the Laplace method requires a transformation of each term of the equation (taking the initial conditions into account) with the subsequent transformed equation then being solved, after which this solution is returned to the time domain by an inverse Laplace transformation. First of all we investigate the general damped single degree of freedom oscillator with an as yet undefined and generalized forcing function and then take Laplace transforms so that we have

$$mL\ddot{x}(t) + cL\dot{x}(t) + kLx(t) = LF(t). \tag{1.113}$$

We then need to transform the derivatives and this gives

$$L\ddot{x}(t) = \int_0^\infty e^{-st} \ddot{x}(t) \, dt = s^2 \bar{x}(s) - sx(0) - \dot{x}(0), \tag{1.114}$$

$$L\dot{x}(t) = \int_0^\infty e^{-st} \dot{x}(t) \, dt = s\bar{x}(s) - x(0), \tag{1.115}$$

where $x(0)$ and $\dot{x}(0)$ are the initial displacement and velocity of the mass m. The Laplace transform of $F(t)$ is simply

$$F(s) = LF(t) = \int_0^\infty e^{-st} F(t) \, dt \tag{1.116}$$

thus, substitution of transformations (1.114), (1.115) and (1.116) into equation (1.113) leads to

$$(ms^2 + cs + k)\bar{x}(s) = \bar{F}(s) + m\dot{x}(0) + (ms + c)x(0). \tag{1.117}$$

If we ignore the homogeneous solution to equation (1.113) so that the initial conditions are zero it is possible to quickly write down the ratio of transformed excitation and transformed response

$$\frac{\bar{F}(s)}{\bar{x}(s)} = ms^2 + cs + k = \frac{1}{\bar{G}(s)}. \tag{1.118}$$

$G(s)$ is the system transfer function and is a concept that also has important consequences in control system design. Rewriting equation (1.118) gives us a definition for the transformed response

$$\bar{x}(s) = \bar{G}(s)\bar{F}(s). \tag{1.119}$$

To obtain the time domain response $x(t)$ we merely need to take the inverse transform of $\bar{x}(s)$, therefore,

$$\bar{x}(t) = L^{-1}\bar{x}(s) = L^{-1}\bar{G}(s)\bar{F}(s). \tag{1.120}$$

Although the inverse transformation operation could take us into the realms of complex line integrals this is almost always avoided in an explicit sense since many standard forms have already been evaluated and are reproduced in the literature. Perhaps the most difficult aspect of applying Laplace transformations is in re-expressing the transformed response $\bar{x}(s)$ into functions which are simple enough to be inverse-transformed using the standard forms; one convenient way of dealing with this is to decompose $\bar{x}(s)$ by means of partial fractions. Having identified and clarified the role of the transfer function we return to equation (1.117) and rewrite it as a sum of the following partial fractions (noting that $c = 2\xi\omega_n$ and $\omega_n^2 = k/m$)

$$\bar{x}(s) = \frac{\bar{F}(s)}{m(s^2 + 2\xi\omega_n s + \omega_n^2)} + \frac{(s + 2\xi\omega_n)x_0}{(s^2 + 2\xi\omega_n s + \omega_n^2)}$$

$$+ \frac{\dot{x}_0}{s^2 + 2\xi\omega_n s + \omega_n^2}. \tag{1.121}$$

The initial conditions are reintroduced here such that $x(0) = x_0$ and $\dot{x}(0) = \dot{x}_0$. At this stage we can begin to look at a specific problem; one which treats a non-periodic pulse type of function defined by

$$F(t) = F_0, \qquad 0 \leqslant t \leqslant T,$$

$$F(t) = 0, \qquad t > T. \tag{1.122}$$

The Laplace transform of the function (1.122) is

$$\bar{F}(s) = LF(t) = \frac{F_0}{s}(1 - e^{-sT}). \tag{1.123}$$

Substitution of the transformation (1.123) into the transformed response (1.121) leads to

$$\bar{x}(s) = \frac{F_0}{m\omega_n^2 s(s^2/\omega_n^2 + 2\xi s/\omega_n + 1)}$$

$$- \frac{F_0 e^{-sT}}{m\omega_n^2 s(s^2/\omega_n^2 + 2\xi s/\omega_n + 1)}$$

$$+ \frac{x_0 s}{\omega_n^2(s^2/\omega_n^2 + 2\xi s/\omega_n + 1)}$$

$$+ \left(\frac{2\xi x_0}{\omega_n} + \frac{\dot{x}_0}{\omega_n^2} \right) \frac{1}{(s^2/\omega_n^2 + 2\xi s/\omega_n + 1)}. \qquad (1.124)$$

Taking inverse transforms and rearranging gives

$$x(t) = \frac{F_0}{m\omega_n^2\sqrt{1 - \xi^2}} \left[-\exp(-\xi\omega_n t). \sin(\omega_n\sqrt{1 - \xi^2}\, t + \arccos \xi) \right.$$

$$\left. + \exp(-\xi\omega_n[t - T]). \sin\{\omega_n\sqrt{1 - \xi^2}\, (t - T) + \arccos \xi\} \right]$$

$$- \frac{x_0}{\sqrt{1 - \xi^2}} \exp(-\xi\omega_n t) \sin .(\omega_n\sqrt{1 - \xi^2}\, t - \arccos \xi)$$

$$+ \frac{(2\xi\omega_n x_0 + \dot{x}_0)}{\omega_n\sqrt{1 - \xi^2}} \exp(-\xi\omega_n t) \sin .(\omega_n\sqrt{1 - \xi^2}\, t). \qquad (1.125)$$

The full damped response of the single degree of freedom oscillator to the rectangular pulse $F(t)$ is given above in equation (1.125). It is obvious that even this method is not entirely free of cumbersome algebra, however on balance it is by far and away the most amenable method available for transient vibration problems like this. The response $x(t)$ with time is depicted in Fig. 1.20.

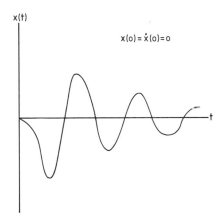

Figure 1.20 Damped response of single degree of freedom oscillator to a single, rectangular, excitational pulse.

1.7 SYSTEMS WITH MORE THAN ONE DEGREE OF FREEDOM

The foregoing text has dealt with the basics of single degree of freedom problems and has briefly considered the free, forced and impulsed categories of vibrational behaviour. In doing this we have consistently regarded the mass or inertia which is in motion as one single, inelastic lump; this does of course limit our scope since a very significant proportion of mechanical vibration problems do not realistically fall into this simplistic category and may contain two or more separate and clearly discernible masses or inertias. Whilst two degree of freedom problems are not, in general, particularly tedious to treat we find that the scale of aglebraic manipulation necessary for multi-degree of freedom demands the formulation of suitable matrix methods as a means of expressing and handling the required set of governing equations.

1.7.1 Using matrices to represent flexibility and stiffness

We can construct the equations of motion for any vibrating system containing at least two degrees of freedom by considering the elastic and inertial aspects of the problem as matrix elements. In the case of system elasticity we can consider it either in terms of flexibility (and use the method of Influence Coefficients) or in terms of stiffnesses, and here it should be noted that each of these quantities is the inverse of the other and that our choice of representation depends largely on our further choice of analysis and indeed the end point we hope to reach.

In general a structure may be loaded by several forces acting at different points such that each point deflects by some amount which is due to the summed contribution of these forces

$$x_{jk} = a_{jk}F_k \tag{1.126}$$

being the displacement at some point j due to a force acting at point k. The influence coefficient is a_{jk}. Therefore by considering all the forces and all the deflections we find that

$$x_1 = a_{11}F_1 + a_{12}F_2 + a_{13}F_3 + \ldots + a_{1n}F_n, \tag{1.127}$$

$$x_2 = a_{21}F_1 + a_{22}F_2 + a_{23}F_3 + \ldots + a_{2n}F_n, \tag{1.128}$$

and so on to x_n. This is expressed more neatly by

$$x_j = \sum_{k=1}^{n} x_{jk} = \sum_{k=1}^{n} a_{jk}F_k \qquad j = 1, 2, \ldots, n. \tag{1.129}$$

In matrix notation such a load/deflection problem is simply

$$\bar{x} = [a]\bar{F}. \tag{1.130}$$

If we examine equation (1.130) more closely and regard the problem as being n-order we have

$$
\begin{bmatrix} x_1 \\ x_2 \\ x_n \end{bmatrix} = \begin{bmatrix} a_{11} & a_{12} & \cdots & a_{1n} \\ a_{21} & a_{22} & \cdots & a_{2n} \\ a_{n1} & a_{n2} & \cdots & a_{nn} \end{bmatrix} \begin{bmatrix} F_1 \\ F_2 \\ F_n \end{bmatrix}. \tag{1.131}
$$

The inverse technique is to define the force at point j as being due to some displacement (usually a unit displacement) at k with all other points fixed. As the other points will not necessarily be fixed if we have multi-point loading, the overall picture will be described by

$$
F_j = \sum k_{kj} x_k, \qquad j = 1, 2, \ldots, n, \tag{1.132}
$$

or,

$$
\bar{F} = [k]\bar{x}, \tag{1.133}
$$

where the bar denotes a vector.

Written in full for an n-order system equation (1.133) becomes

$$
\begin{bmatrix} F_1 \\ F_2 \\ F_n \end{bmatrix} = \begin{bmatrix} k_{11} & k_{12} & \cdots & k_{1n} \\ k_{21} & k_{22} & \cdots & k_{2n} \\ k_{n1} & k_{n2} & \cdots & k_{nn} \end{bmatrix} \begin{bmatrix} x_1 \\ x_2 \\ x_n \end{bmatrix}. \tag{1.134}
$$

The inverse relationship between $[a]$ and $[k]$ is relatively easy to prove; if we take equation (1.130) and premultiply by $[a]^{-1}$ we get the following

$$
[a]^{-1}\bar{x} = [a]^{-1}[a]\,\bar{F}, \tag{1.135}
$$

$$
\therefore\ [a]^{-1}\bar{x} = \bar{F}. \tag{1.136}
$$

However we have established that

$$
\bar{F} = [k]\bar{x}.
$$

Thus

$$
[a]^{-1} = [k] \tag{1.137}
$$

and hence

$$
[a] = [k]^{-1}. \tag{1.138}
$$

One can also show that the rule of matrix symmetry applies such that $a_{jk} = a_{kj}$ and $k_{jk} = k_{kj}$, by means of the theorem of reciprocity.

1.7.2 The eigenvalue problem

The single degree of freedom configuration is expressed at its most simplest by equation (1.10) in which we make no allowances for excitational and dissipatory forces acting on and in the system. This simplified representation also has a relevance to multi-degree of freedom systems in that we can restate equation (1.10) in matrix form,

$$[m]\ddot{x} + [k]x = 0. \tag{1.139}$$

By assuming a harmonic solution form

$$\bar{x} = \bar{X}\sin\omega t \tag{1.140}$$

and rearranging, equation (1.139) becomes

$$[k]\bar{X} = \omega^2[m]\bar{X}. \tag{1.141}$$

Premultiplication by $[m]^{-1}$ leads to

$$[m]^{-1}[k]\bar{X} = [m]^{-1}[m]\omega^2\bar{X}. \tag{1.142}$$

The product $[m]^{-1}[k]$ is usually referred to as the dynamic matrix $[A]$. Rearranging and rewritting equation (1.142) yields the eigenvalue problem formulation

$$[A - \omega^2 I]\bar{X} = 0, \tag{1.143}$$

where $[I]$ is simply a unit matrix. Conventionally one writes ω^2 as λ and we can proceed to find λ values which will satisfy equation (1.143) by considering the determinant of the matrix. In this way we investigate the following characteristic equation.

$$|A - \lambda I| = 0 \tag{1.144}$$

for which the roots, λ_i, are known as eigenvalues. By substituting the λ_i into equation (1.143) we can calculate the mode shape or eigenvector corresponding to each λ_i. Therefore an n-degree of freedom system will have n eigenvalues and n eigenvectors. The characteristic equation (1.144) is expanded into an n-order polynomial in λ. This is solved by straightforward analysis up to quadratic order or by zero crossing iteration for order 3 and above.

1.7.3 Nonconservative terms, modal analysis

If we first of all consider the case of an external force on an n-degree of freedom system we would naturally assume that modelling of the problem would generate n coupled differential equations of motion. Solution of the problem as described is frustrated by the sheer complexity introduced by a large number of degrees of freedom, and so it is of considerable importance that we reflect on system decoupling; a process whereby we restate the problem in the form of n decoupled equations, each one solvable on its own. This approach is called **modal analysis**.

Equation (1.139) is extended to include a general and arbitrary external forces vector F

$$[m]\ddot{\bar{x}} + [k]\bar{x} = \bar{F}. \tag{1.145}$$

Initially one invokes the eigenvalue form of the conservative problem, equation (1.141), in order to establish the linear undamped natural frequencies $\omega_{1,2,...,n}$ and their corresponding normal modes $\bar{X}_{1,2,...,n}$. The normal modes (or eigenvectors), $\bar{X}_{1,2,...,n}$ can be combined in the form of an additive linear series

$$\bar{x} = q_1(t)\bar{X}_1 + q_2(t)\bar{X}_2 + \ldots + q_n(t)\bar{X}_n, \tag{1.146}$$

where the $q_i(t)$ are time dependent generalized modal coordinates. Incorporation of the $q_i(t)$ and the X_i into a modal vector and modal matrix, respectively, allows us to rewrite equation (1.146) as

$$\bar{x} = [X]\bar{q}. \tag{1.147}$$

The eigenvectors X_i are purely spatial functions and so substitution into equation (1.145), after premultiplication by the modal matrix transpose, leads to

$$[X]^{\mathrm{T}}[m][X]\ddot{\bar{q}} + [X]^{\mathrm{T}}[k][X]\bar{q} = [X]^{\mathrm{T}}\bar{F}. \tag{1.148}$$

In order to complete the decoupling of the system we need to introduce the property of modal orthogonality. At its simplest we can define mutual orthogonality of eigenvectors as a result of the scalar product being shown to be equal to zero. If we return to equation (1.141) and identify the two eigenvalues and eigenvectors which relate to the ith and jth modes we have

$$[k]\bar{X}^{(i)} = \omega_i^2[m]\bar{X}^{(i)} \tag{1.149}$$

and

$$[k]\bar{X}^{(j)} = \omega_j^2[m]\bar{X}^{(j)}. \tag{1.150}$$

Premultiplication of equations (1.149) and (1.150) by transposes $X^{(j)\mathrm{T}}$ and $X^{(i)\mathrm{T}}$ in turn produces

$$\bar{X}^{(j)\mathrm{T}}[k]\bar{X}^{(i)} = \omega_i^2 \bar{X}^{(j)\mathrm{T}}[m]\bar{X}^{(i)} \tag{1.151}$$

and

$$\bar{X}^{(i)\mathrm{T}}[k]\bar{X}^{(j)} = \omega_j^2 \bar{X}^{(i)\mathrm{T}}[m]\bar{X}^{(j)}. \tag{1.152}$$

If matrices $[m]$ and $[k]$ are symmetrical then we could interchange the i and j on both sides of equation (1.151) (say) to get

$$\bar{X}^{(i)\mathrm{T}}[k]\bar{X}^{(j)} = \omega_i^2 \bar{X}^{(i)\mathrm{T}}[m]\bar{X}^{(j)}. \tag{1.153}$$

Subtraction of equation (1.152) from (1.153) leads to

$$(\omega_i^2 - \omega_j^2)\bar{X}^{(i)\mathrm{T}}[m]\bar{X}^{(j)} = 0. \qquad (1.154)$$

We can be generally confident that $\omega_i \neq \omega_j$ (except for the special case of repeated eigenvalues where $\omega_i = \omega_j$ for which the relevant eigenvectors cannot be mutually orthogonal, however this does not prevent them being orthogonal to other modal vectors), therefore,

$$\bar{X}^{(i)\mathrm{T}}[m]\bar{X}^{(j)} = 0. \qquad (1.155)$$

Obviously we can also work on the stiffness matrix, and this gives

$$\bar{X}^{(i)\mathrm{T}}[k]\bar{X}^{(j)} = 0. \qquad (1.156)$$

Both equations require $i \neq j$.

Equations (1.155) and (1.156) are conclusive statements of orthogonality of normal modes i and j since it is quite clear that the scalar product in the context both of mass and stiffness equals zero.

We proceed to use equations (1.155) and (1.156) in the decoupling of the generalized problem defined by equation (1.145) by setting $i = j$ such that the bracketed term of equation (1.154) goes to zero, leaving us with

$$[X]^{\mathrm{T}}[m][X] \neq 0 = [M], \qquad (1.157)$$

$$[X]^{\mathrm{T}}[k][X] \neq 0 = [K], \qquad (1.158)$$

for $i = 1, 2, \ldots, n$, meaning that $[M]$ and $[K]$ respectively denote the generalized mass and stiffness for the ith mode, whichever one that may be, and that these matrices are therefore diagonal with M_{rr} and K_{rr} as the elements. The eigenvector is now formally rewritten as a modal matrix $[X]$, which can be orthonormalized so that $M_{11} = M_{22} = M_{rr} = 1$ and so we get

$$\bar{X}^{(i)\mathrm{T}}[m]\bar{X}^{(i)} = 1, \qquad \text{for} \quad i = 1, 2, \ldots, n, \qquad (1.159)$$

and thus

$$[X]^{\mathrm{T}}[m][X] = [M] = \lceil 1 \rfloor \qquad (1.160)$$

the unit matrix. From this we can show that

$$[X]^{\mathrm{T}}[k][X] = [K] = \lceil \omega^2 \rfloor. \qquad (1.161)$$

Equations (1.160) and (1.161) are now substituted into equation (1.148)

$$\lceil 1 \rfloor \ddot{q} + \lceil \omega^2 \rfloor q = [X]^{\mathrm{T}} \bar{F}. \qquad (1.162)$$

The right-hand side of equation (1.162) may be expressed as a generalized forces vector Q which is specifically associated with the generalized coordinates q.

The result is simply

$$\ddot{q}_i + \omega_i^2 q_i = Q_i, \qquad i = 1, 2, \ldots, n, \qquad (1.163)$$

a set of n uncoupled equations of motion.

It can be seen that equations (1.163) describe forced undamped vibration in each mode through $i = 1, 2, \ldots, n$ and that they can be solved independently of one another. The convolution integral is a useful concept here and is reintroduced so that the particular solution due to the arbitrary Q_i is

$$q_i^{(\mathrm{p})} = \int_0^t Q_i(\tau) h_i(t - \tau) \, d\tau, \qquad (1.164)$$

where $h_i(t - \tau)$ is the undamped time shifted impulse response such that

$$h_i(t - \tau) = \frac{1}{\omega_i} \sin \omega_i (t - \tau). \qquad (1.165)$$

Substitution of equation (1.165) into (1.164) produces the particular solution

$$q^{(\mathrm{p})} = \frac{1}{\omega_i} \int_0^t Q_i(\tau) \sin \omega_i (t - \tau) \, d\tau \qquad \text{for all } i. \qquad (1.166)$$

The complementary solution $q_i^{(\mathrm{c})}$ is obtained from equation (1.15) thus,

$$q_i^{(\mathrm{c})} = q_i(0) \cos \omega_i t + \frac{\dot{q}_i(0)}{\omega_i} \sin \omega_i t. \qquad (1.167)$$

The full solution is

$$q_i^{(\mathrm{F})}(t) = q_i^{(\mathrm{c})} + q_i^{(\mathrm{p})}. \qquad (1.168)$$

Therefore,

$$q_i^{(\mathrm{F})}(t) = q_i(0) \cos \omega_i t + \frac{\dot{q}_i(0)}{\omega_i} \sin \omega_i t + \frac{Q_i(\tau)}{\omega_i^2} (1 - \cos \omega_i t). \quad (1.169)$$

Superposition of the modal reponses enables us to construct the solution to the original equation (1.145) by virtue of equation (1.147), so that we obtain

$$\bar{x} = \sum_{i=1}^{n} X_i q^{(\mathrm{F})}(t). \qquad (1.170)$$

It has already been pointed out that the eigenfunctions can be derived with recourse to equation (1.139) and the subsequent analysis. This is most conveniently portrayed in the form of a numerical example.

Equation (1.139) represents a multi-degree of freedom system and in order to demonstrate the technique of eigenfunction calculations we will restrict it to three degrees of freedom with the following mass and stiffness matrices (which could characteristically model a three degree of freedom fixed–free shaft–rotor configuration, for example)

$$[m] = \begin{bmatrix} 2 & 0 & 0 \\ 0 & 1.25 & 0 \\ 0 & 0 & 1 \end{bmatrix};$$

$$[k] = \begin{bmatrix} 3.8 & -2 & 0 \\ -2 & 3.5 & -1.5 \\ 0 & -1.5 & 1.5 \end{bmatrix}. \tag{1.171}$$

The dynamic matrix

$$[A] = [m]^{-1}[k] = \begin{bmatrix} 1.9 & -1.0 & 0 \\ -1.6 & 2.8 & -1.2 \\ 0 & -1.5 & 1.5 \end{bmatrix}. \tag{1.172}$$

Clearly we write

$$[A - \lambda I] = \begin{bmatrix} (1.9 - \lambda) & -1 & 0 \\ -1.6 & (2.8 - \lambda) & -1.2 \\ 0 & -1.5 & (1.5 - \lambda) \end{bmatrix} \tag{1.173}$$

from equation (1.143).

The characteristic equation is simply the determinant of equation (1.173) equated to zero and we can quite easily expand this 3×3 determinant to arrive at the characteristic polynomial in λ. This is,

$$-\lambda^3 + 6.2\lambda^2 - 8.97\lambda + 2.16 = 0 \tag{1.174}$$

with roots $\lambda_1 = 0.300$; $\lambda_2 = 1.725$; $\lambda_3 = 4.175$.

In a stiffness matrix based problem such as this we have

$$\lambda = \omega_i^2 \tag{1.175}$$

so the three undamped natural frequencies are

$$\omega_1 = 0.547; \qquad \omega_2 = 1.313; \qquad \omega_3 = 2.043.$$

Having established the natural frequencies it is necessary to evaluate the eigenvectors; this is achieved by writing equation (1.143) in full

$$\begin{bmatrix} (1.9 - \lambda_i) & -1 & 0 \\ -1.6 & (2.8 - \lambda_i) & -1.2 \\ 0 & -1.5 & (1.5 - \lambda_i) \end{bmatrix} \begin{bmatrix} X_1 \\ X_2 \\ X_3 \end{bmatrix}_i = 0. \tag{1.176}$$

First row expansion leads to

$$\left(\frac{X_2}{X_1}\right)_i = (1.9 - \lambda_i). \tag{1.177}$$

Third row expansion gives

$$\left(\frac{X_3}{X_2}\right)_i = \frac{1.5}{1.5 - \lambda_i} \tag{1.178}$$

since there are three unknowns and only two equations we normalize one 'end' of the eigenvectors, say $X_{3i} = 1$, giving

$$X_{1i} = \frac{1.5 - \lambda}{1.5(1.9 - \lambda)},$$ (1.179)

$$X_{2i} = \frac{1.5 - \lambda_i}{1.5}.$$ (1.180)

It should be noted that the second row is not expanded because it would give no new information. Substitution of λ_i into X_{1i}, X_{2i}, X_{3i} gives the normalized eigenvectors for the three modes, therefore

$$X_1 = \begin{bmatrix} 0.5 \\ 0.8 \\ 1 \end{bmatrix}; \quad X_2 = \begin{bmatrix} -0.85 \\ -0.15 \\ 1 \end{bmatrix}; \quad X_3 = \begin{bmatrix} 0.78 \\ -1.73 \\ 1 \end{bmatrix}.$$ (1.181)

These are depicted in Fig. 1.21. The eigenvectors (or mode shapes) could be substituted into equation (1.170) so that a full solution for x is obtained. We would need to explicitly define F, and hence Q_i, and would also have to know $q_i(0)$ and $\dot{q}_i(0)$ (which might well both be zero), before the full solution could be completely determined. It will be appreciated that this technique could be applied to systems with many more degrees of freedom.

The next level of difficulty involves the inclusion of viscous damping, and we find here that there will be severe physical constraints on the damping characteristics and/or magnitude to ensure that the modal damping matrix is diagonal. An analytical method is available for

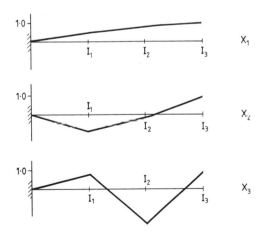

Figure 1.21 Normalized eigenvectors for a three degree of freedom fixed–free shaft–rotor assembly.

non-diagonal modal damping matrix problems (Meirovitch, 1986) which introduces the concept of complex eigenfunctions. It is possible to consider a restricted range of cases in which the damping matrix can be regarded as a linear additive combination of the mass and stiffness matrices (Rayleigh, 1945), so that,

$$[c] = a[m] + b[k] + \ldots . \tag{1.182}$$

This will be reasonably accurate as long as the damping is relatively light, and in practice such a model will enable us to tend towards a diagonal damping matrix. On this basis any non-zero non-diagonal terms will be of significantly smaller magnitude than those of the diagonal and can therefore reasonably be neglected. It can be shown (Warburton, 1976) that terms additional to those of equation (1.182) may be inserted if equation (1.141) is premultiplied by $[X]^T[K][M]^{-1}$ or alternatively by $[X]^T[M][K]^{-1}$ which would each give rise to additional orthogonality relations, and hence additional terms in equation (1.182). However, as corroborated by Warburton, further extension to equation (1.182) would not normally be necessary in most practical situations where the damping is fairly light. The complex eigenfunction method of Meirovitch is appropriate to problems in which the damping is too heavy to permit this approximation. It will not be dealt with here.

Assuming that we are dealing with a system for which equation (1.182) holds and in which the actual magnitude of the damping is quite low equation (1.145) can be extended to incorporate this

$$[m]\bar{\bar{x}} + [c]\bar{y} + [k]\bar{x} = \bar{F}. \tag{1.183}$$

Equation (1.148) would therefore be of the form

$$[X]^T[m][X]\bar{\bar{q}} + [X]^T[c][X]\bar{q} + [X]^T[k][X]\bar{q} = [X]^T\bar{F} \tag{1.184}$$

after substitution of equation (1.147) and premultiplication by transpose $[X]^T$. Substitution of equation (1.182) into (1.184) gives

$$[X]^T[m][X]q + [a[X]^T[m][X] + b[X]^T[k][X]]\bar{q}$$
$$+ [X]^T[k][X]\bar{q} = [X]^T\bar{F}. \tag{1.185}$$

Substitution of equations (1.160) and (1.161) into (1.185) leads to

$$[I]\bar{\bar{q}} + [a[I] + b[\omega^2]]\bar{q} + [\omega]^2\bar{q} = [X]^T\bar{F}. \tag{1.186}$$

Equation (1.186) is written as a set of n decoupled equations of motion, analogous to those of (1.163),

$$\ddot{q}_i + (a + \omega_i^2 b)\dot{q}_i + \omega_i^2 q_i = Q_i, \qquad i = 1, 2, \ldots, n. \tag{1.187}$$

Solution of this type of equation is covered in Section 1.6.5, and we could set up the full solution to equation (1.183) by utilizing (1.170) and then by calculating the appropriate eigenfunctions. It is quite appropriate to involve the calculated eigenvectors (or normal modes) of the undamped system as long as the damped system of interest has classical normal modes which is indeed the case provided Rayleigh's proposed form of damping in equation (1.182) is in force. The use of Rayleigh's form is investigated in the literature (Caughey, 1960).

As the number of degrees of freedom is increased the complexity and tedium of algebraic manipulations and numerical eigenfunction calculations could outweigh the usefulness of the desired result, and so for cases like this computer software has been developed to cater for large scale discretized problems. It should also be noted that it may be easier to calculate the modal eigenfunctions by the method of matrix iteration, a technique which obviates the necessity for setting up and solving a characteristic equation. Iterative techniques are eminently suitable for computerization. In addition to this there are several extremely powerful commercial modal analysis packages available which are based on finite element models of large scale discretized dynamical problems, in particular the Entek EMODAL 30 package (Hewlett–Packard compatible) and the SMS (Scientific Atlanta) 5005A Starmodal software which runs on IBM equipment.

1.7.4 Virtual work and Lagrange's equations as applied to multi-degree of freedom problems

Most practitioners of vibrational analysis develop a ready facility with Newtonian modelling of simple problems, however this approach tends to become less suitable as the number of degrees of freedom increases, and also when the system masses or inertias are not straightforwardly interconnected by the stiffnesses and dissipative elements present in the system. In cases such as these it is easier to proceed along the non-vectorial lines of energy and work, and to this end the important (and complementary) theories of virtual work and Lagrange's formulation are to be developed.

Section 1.1.1 discussed the classification of dynamical systems in terms of the constraints acting and it was pointed out that coordinates that equated in quantity to the number of degrees of freedom of the system (and were therefore not present in any constraint equations that might apply) were known to be independent or generalized coordinates. This is an important basis for the work of this section.

Initially we consider a virtual displacement; this should be regarded as a minute positional change in a coordinate in any direction, as long as it does not violate the constraints that are in force. The time involved is irrelevant, and in order to develop the theory we must conjecture that we are dealing with some arbitrary system of particles on which a selection of forces are acting such that the overall system (and each individual particle) is in static equilibrium. This means that the resultant force $\bar{R}_j$ on each and every particle j is zero and that the work done on the particle through a virtual displacement $\delta \bar{r}_j$ is therefore zero.

Then we will have

$$\delta W = \sum \bar{R}_j \delta \bar{r}_j = 0 \qquad (1.188)$$

from our definition of work given in Section 1.2.1.

$\bar{R}_j$ will contain both applied and constraint forces, $\bar{F}_j$ and $\bar{f}_j$, and we are assuming that the constraint forces do not do any work, therefore,

$$\delta W = \sum_j \bar{F}_j \delta \bar{r}_j = 0. \qquad (1.189)$$

The arbitrary displacement $\bar{r}_j$ may be regarded as a function of independent, generalized, coordinates (up to n), and time too since $q_{1 \to n}$ are time dependent, and so we take an n degree of freedom system, giving

$$\bar{r}_j = \bar{r}_j(q_1, q_2, \ldots, q_n, t). \qquad (1.190)$$

Taking the ith coordinate and concerning ourselves purely with the spatial aspect of the problem it is possible to write

$$\delta r_j = \sum_i \frac{\partial \bar{r}_j}{\partial q_i} \delta q_i. \qquad (1.191)$$

Substituting this into equation (1.189) gives

$$\delta W = \sum_j \sum_i \bar{F}_j \frac{\partial \bar{r}_j}{\partial q_i} \delta q_i \qquad (1.192)$$

for which it is possible to define a generalized force of the form

$$Q_i = \sum_j \bar{F}_j \frac{\partial \bar{r}_j}{\partial q_i}. \qquad (1.193)$$

Substitution of equation (1.193) into (1.192) leads to the well-known expression for virtual work of a system in terms of its generalized coordinates

$$\delta W = \sum_{i=1}^{n} Q_i \delta q_i. \qquad (1.194)$$

This generalized approach, whilst not being too rigorous, allows us to progress to developing the Lagrangian formulation both for conservative and non-conservative systems. Once derived it is relatively easy to apply the relevant form of Lagrange's equations to a specific vibration problem.

A particle under the influence of a force system will accelerate, however dynamic equilibrium is created if an equal and opposite force is introduced (D'Alembert's Principle), so by considering again the applied and constraint forces $\bar{F}_i$ and $\bar{f}_i$ we state this in the following form

$$\bar{F}_i + \bar{f}_i - m_i\bar{\ddot{r}}_i = 0. \tag{1.195}$$

The constraint force can once again be discounted and so the virtual work for the whole system must be

$$\sum_i (\bar{F}_i - m_i\bar{\ddot{r}}_i)\delta\bar{r}_i = 0. \tag{1.196}$$

The velocity vector $\bar{\dot{r}}_i$ is derived from equation (1.190),

$$\bar{\dot{r}}_i = \frac{\partial\bar{r}_i}{\partial q_1}\dot{q}_i + \frac{\partial\bar{r}_i}{\partial q_2}\dot{q}_2 + \ldots + \frac{\partial\bar{r}_i}{\partial q_n}\dot{q}_n + \frac{\partial\bar{r}_i}{\partial t}. \tag{1.197}$$

Therefore,

$$\frac{\partial\bar{\dot{r}}_i}{\partial\dot{q}_k} = \frac{\partial\bar{r}_i}{\partial q_k}, \tag{1.198}$$

where q_k is the kth generalized coordinate.

Again from consideration of equation (1.191) we can show that the virtual displacement of $\bar{r}_i$ is

$$\delta\bar{r}_i = \frac{\partial\bar{r}_i}{\partial q_1}\delta q_1 + \frac{\partial\bar{r}_i}{\partial q_2}\delta q_2 + \ldots + \frac{\partial\bar{r}_i}{\partial q_n}\delta q_n$$

$$= \sum_{k=1}^{n} \frac{\partial\bar{r}_i}{\partial q_k}\delta q_k. \tag{1.199}$$

Equation (1.196) can be expanded and equation (1.199) substituted into the second term leading to

$$m_i\bar{\ddot{r}}_i\delta\bar{r}_i = \sum_{k=1}^{n} m_i\bar{\ddot{r}}_i \frac{\partial\bar{r}_i}{\partial q_k}\delta q_k. \tag{1.200}$$

By considering one term of the summation on the right-hand side of equation (1.200) we can write the following

$$m_i\bar{\ddot{r}}_i\frac{\partial\bar{r}_i}{\partial q_k} = \frac{d}{dt}\left(m_i\bar{\dot{r}}_i\frac{\partial\bar{r}_i}{\partial q_k}\right) - m_i\bar{\dot{r}}_i\frac{d}{dt}\left(\frac{\partial\bar{r}_i}{\partial q_k}\right). \tag{1.201}$$

Substituting equation (1.198) enables equation (1.201) to be expressed in the form below (in which it can be seen that kinetic energy is explicit) so that

$$m_i\bar{\ddot{r}}_i\delta\bar{r}_i = \sum_{k=1}^{n}\left[\frac{d}{dt}\frac{\partial}{\partial\dot{q}_k} - \frac{\partial}{\partial q_k}\right](\tfrac{1}{2}m_i\dot{r}_i^2)\delta q_k. \tag{1.202}$$

Summing over i particles with the kinetic energy defined as

$$T = \tfrac{1}{2} \sum m_i \dot{r}_i^2,$$ (1.203)

$$\sum_i m_i \bar{\dot{r}}_i \delta \bar{r}_i = \sum_{k=1}^{n} \left[\frac{\mathrm{d}}{\mathrm{d}t} \frac{\partial}{\partial \dot{q}_k} - \frac{\partial}{\partial q_k} \right] \delta q_k.$$ (1.204)

From equations (1.194) and (1.193) the virtual work associated with coordinate q_k is

$$\delta W = \sum_{k=1}^{n} Q_k \delta q_k = \sum \left(\sum \bar{F}_i \frac{\partial \bar{r}_i}{\partial q_k} \right) \delta q_k$$ (1.205)

allowing us to write from equation (1.196)

$$\sum_{k=1}^{n} \left[\frac{\mathrm{d}}{\mathrm{d}t} \left(\frac{\partial T}{\partial \dot{q}_k} \right) - \frac{\partial T}{\partial q_k} - Q_k \right] \delta q_k = 0.$$ (1.206)

So for the kth coordinate one obtains

$$\frac{\mathrm{d}}{\mathrm{d}t} \left(\frac{\partial T}{\partial \dot{q}_k} \right) - \frac{\partial T}{\partial q_k} - Q_k = 0.$$ (1.207)

For the most useful case which deals with the non-conservative problem the simple constant energy equation (1.16) does not hold and so we introduce a non-conservative energy term defined as the Lagrangian L,

$$L = T - U.$$ (1.208)

We rewrite equation (1.207) substituting L for T to get

$$\frac{\mathrm{d}}{\mathrm{d}t} \left(\frac{\partial L}{\partial \dot{q}_k} \right) - \frac{\partial L}{\partial q_k} = \widetilde{Q}_k,$$ (1.209)

where $\widetilde{Q}_k$ represents all the non-conservative forces present (excitational and dissipative). Using equation (1.208) this becomes

$$\frac{\mathrm{d}}{\mathrm{d}t} \left(\frac{\partial T}{\partial \dot{q}_k} \right) - \frac{\partial T}{\partial q_k} + \frac{\partial U}{\partial q_k} = \widetilde{Q}_k.$$ (1.210)

$\widetilde{Q}_k$ can be defined more explicitly and to do this we interpose the Rayleigh dissipation function

$$R_k = \tfrac{1}{2} c \dot{x}^2.$$ (1.211)

c can, if required, be a (positive definite) matrix describing the system viscous damping as appropriate to a multi-degree of freedom system. So,

$$\widetilde{Q}_k = P_k - R_k.$$ (1.212)

The full and most practically useful form of Lagrange's equations is therefore

$$\frac{\mathrm{d}}{\mathrm{d}t} \left(\frac{\partial T}{\partial \dot{q}_k} \right) - \frac{\partial T}{\partial q_k} + \frac{\partial R}{\partial \dot{q}_k} + \frac{\partial U}{\partial q_k} = P_k.$$ (1.213)

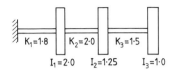

Figure 1.22 Fixed–free shaft–rotor assembly with three lumped inertias (three degrees of freedom).

This is applicable to all vibration problems where a viscous damping model is acceptable. For non-viscous cases one would neglect the $\partial R/\partial \dot{q}_k$ term and would consider the damping mechanism by using an identification method. The terms could then be added on to the derived equation(s) of motion, or R_k could possibly be redefined and the Lagrangian analysis used in full. The mass and stiffness matrices (1.71) can be obtained for the system of Fig. 1.22 by using Lagrange's equations. As the problem is one of negligible damping and free vibration we state

$$\frac{\partial R}{\partial \dot{q}_k} = P_k = 0. \tag{1.214}$$

This simplifies equation (1.213) and merely requires us to write down the K.E. and P.E. expressions for the configuration. Each rotor is assigned one generalized coordinate (representing torsional motion); K.E. and P.E. are, respectively,

$$T = \tfrac{1}{2}I_1\dot{q}_1{}^2 + \tfrac{1}{2}I_2\dot{q}_2{}^2 + \tfrac{1}{2}I_3\dot{q}_3{}^2 = \dot{q}_1{}^2 + 0.625\dot{q}_2{}^2 + 0.5\dot{q}_3{}^2, \tag{1.215}$$

$$U = \tfrac{1}{2}k_1q_1{}^2 + \tfrac{1}{2}k_2(q_2 - q_1)^2 + \tfrac{1}{2}k_3(q_3 - q_2)^2. \tag{1.216}$$

U is therefore

$$U = 1.9q_1{}^2 + 1.75q_2{}^2 + 0.75q_3{}^2 - 2q_1q_2 - 1.5q_2q_3. \tag{1.217}$$

Having three generalized coordinates means that there will be three equations of motion, and so for the first we let $k = 1$ in equations (1.213) and operate on each term successively.

Therefore

$$\frac{\mathrm{d}}{\mathrm{d}t}\left(\frac{\partial T}{\partial \dot{q}_1}\right) = 2\ddot{q}; \qquad \frac{\partial T}{\partial q_1} = 0; \qquad \frac{\partial U}{\partial q_1} = 3.8q_1 - 2q_2. \tag{1.218}$$

This gives us the first equation

$$2\ddot{q}_1 + 3.8q_1 - 2q_2 = 0. \tag{1.219}$$

Similarly for $k = 2$, $k = 3$,

$$1.25\ddot{q}_2 + 3.5q_2 - 2q_1 - 1.5q_3 = 0, \tag{1.220}$$

$$\ddot{q}_3 + 1.5q_3 - 1.5q_2 = 0. \tag{1.221}$$

It can be clearly seen how matrices (1.171) are constructed from these equations.

1.8 CONTINUOUS SYSTEMS

We now move on to a form of analysis that considers the vibrating system as a single entity – a continuous system – thus the concept of separable masses and individual elastic elements is no longer appropriate. The continuous representation treats the system in question as one with an infinite number of degrees of freedom rather than a finite number of identifiable degrees of freedom, as in the discrete model. There are no clearly distinguishable system types that fall into either the discrete or the continuous category, rather it is a matter of desired accuracy as to which is adopted. Partial differential equations are found to govern continuous systems, and so the attendant complexity of solution must be considered alongside the potential gain in accuracy. It is important to realize that a continuous model will generate an infinite number of normal modes and natural frequencies, however, in order to calculate these we will need to explicitly consider the boundary conditions that are in force.

Meirovitch (1986), makes the important assertion that the continuous representation is actually the limiting case for the discrete version, and on this basis he then proceeds to treat the well-known string problem undergoing transverse vibration. The conclusion to this carefully worked example is that the crossover from discrete to continuous representation occurs through the limit of separable lumped masses to a smooth and inseparable continuous mass throughout the string.

In addition to this we find that continuous systems analysis readily admits the concepts of eigenfunctions and orthogonality of normal modes. The conceptual modifications required here are that a continuously represented problem will exhibit an infinite quantity, or set, of eigenvalues and space dependent eigenvectors.

The usual approach to continuous analysis is to explore the examples of transversely vibrating strings, axially vibrating rods, bending of beams, and shafts undergoing torsional vibration. The governing equations are second order with the exception of the bending beam problem which is fourth order. The simple wave equation is invariably cited as a practical form of the string problem for the simplified case of uniform string material, constant tension, and small amplitude motion.

1.8.1 The lateral vibration of uniform beams

The clearest way of obtaining the fourth-order differential equation of motion for general uniform beam problems is to consider an elemental

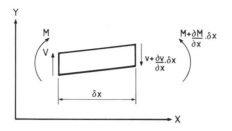

Figure 1.23 Elemental length of a beam undergoing lateral vibration.

length of beam as it undergoes vibratory motion. The forces and moments that occur on and within the beam may then be dealt with more simply. Such an element is depicted in Fig. 1.23. There are several assumptions which have to be made, specifically these are that the length of the beam is very much larger than either of the cross-sectional dimensions and that we can relate the stress distribution through the cross-section to the deformation of the beam. This latter assumption allows us to calculate the moment at any section, therefore,

$$EIy'' = -M, \tag{1.222}$$

where EI is the flexural rigidity, y is the displacement and M is the acting moment at the section. The primes denote differentiation with respect to the axial coordinate x. Also we can sum the vertical forces acting on the element and from this show that

$$V' = \rho A \ddot{y}, \tag{1.223}$$

where ρ is the density, A the elemental cross-section area and V is the shear force, and the dots denote differentiation with respect to time; thus the right-hand side is derived from the elemental inertia force.

Since $V = M'$, we get

$$M'' = \rho A \ddot{y}. \tag{1.224}$$

Therefore substitution of equation (1.224) into (1.222) leads to

$$EIy'''' + \rho A \ddot{y} = 0. \tag{1.225}$$

This is sometimes called the dispersive wave equation. Conventionally we take a free vibration type of solution of the general form

$$y(x, t) = \phi(x) \sin \omega t, \tag{1.226}$$

which, when substituted into equation (1.225) leads to

$$\phi'''' - \beta^4 \phi = 0, \tag{1.227}$$

where

$$\beta^4 = \frac{\omega^2 \rho A}{EI}.$$ (1.228)

The general solution to equation (1.227) is given by the spatial function

$$\phi(x) = A\cos\beta x + B\sin\beta x + C\cosh\beta x + D\sinh\beta x.$$ (1.229)

1.8.2 Boundary conditions for the uniform beam problem

There are four discernible boundary conditions for a beam:

pinned joint $y = 0,$ $M = 0 \Rightarrow \phi = 0,$ $\phi'' = 0$ (1.230)

clamped end $y = 0,$ $y' = 0 \Rightarrow \phi = 0,$ $\phi' = 0$ (1.231)

sliding joint $y' = 0,$ $V = 0 \Rightarrow \phi' = 0,$ $\phi''' = 0$ (1.232)

free end $M = 0,$ $V = 0 \Rightarrow \phi'' = 0,$ $\phi''' = 0$ (1.233)

So, if we know the boundary condition at two points along the length of the beam we will have sufficient information to determine three of the arbitrary constants in terms of the fourth of equation (1.229), $(A–D)$, plus a frequency equation. For example, a cantilever of length l would have the following four boundary conditions:

At $x = 0,$ $y = 0,$ $y' = 0$ and $\phi = 0,$ $\phi' = 0,$

At $x = l,$ $M = 0,$ $V = 0$ and $\phi'' = 0,$ $\phi''' = 0.$

EXAMPLE

Pinned–pinned beam:
   The boundary conditions for this configuration are substituted into the general solution (1.229) to give

$$\phi(x) = A\cos\beta x + B\sin\beta x + C\cosh\beta x + D\sinh\beta x,$$
$$\phi''(x) = -A\beta^2\cos\beta x - B\beta^2\sin\beta x + C\beta^2\cosh\beta x + D\beta^2\sinh\beta x.$$ (1.234)

(i) $x = 0$

This leads to

$$0 = A + C,$$

$$\text{i.e. } A = C = 0.$$ (1.235)

$$0 = -A + C,$$

(ii) $x = l$

$$0 = \quad B \sin \beta l + D \sinh \beta l, \tag{1.236}$$

$$0 = -B\beta^2 \sin \beta l + D\beta^2 \sinh \beta l, \qquad \beta^2 \neq 0. \tag{1.237}$$

Equations (1.236) and (1.237) can be added to result in

$$2D \sinh \lambda l = 0 \tag{1.238}$$

from which we find that $D = 0$ for a non-trivial solution. Therefore from equation (1.236)

$$B \sin \beta l = 0, \qquad \sin \beta l = 0, \tag{1.239}$$

which leads to

$$\beta l = \pi, 2\pi, 3\pi, \ldots, n\pi. \tag{1.240}$$

Equation (1.228) defined β, thus we can state βl,

$$\beta l = \sqrt{\frac{\omega^2 \rho A}{EI}}\, l = n\pi \qquad \text{where} \quad n = 1, 2, 3, \ldots. \tag{1.241}$$

Thus,

$$(n\pi)^2 = \sqrt{\frac{\rho A}{EI}}\, l^2 \omega \tag{1.242}$$

leading to

$$\omega = \frac{n^2 \pi^2}{l^2} \sqrt{\frac{EI}{\rho A}}. \tag{1.243}$$

For the fundamental, $n = 1$, so

$$\omega_1 = \frac{EI}{\rho A} \tag{1.244}$$

and, since $A = C = D = 0$ we have the following for the first mode shape,

$$\phi_1(x) = B \sin \beta x \qquad \text{for which} \quad \beta = \sqrt{\frac{\pi}{l}} \qquad \text{(from (1.240))}$$

so

$$\phi_1(x) = B \sin\left(\frac{\pi x}{l}\right). \tag{1.245}$$

1.8.3 Torsional oscillations of shafts

In Section 1.8 it was stated that the one-dimensional wave equation is appropriate to the string problem; perhaps a more common and practical engineering use for this is in the analysis of shafts in torsion. As for the lateral beam vibration problem of Section 1.8.1 we take an

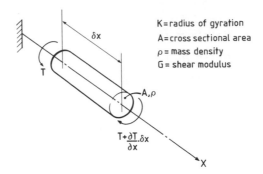

Figure 1.24 Elemental length of a shaft in torsion.

elemental length of the component, in this case a length dx of shaft as shown in Fig. 1.24. The element is twisted through an angle $\delta\theta$ such that simple torsion theory for a uniform shaft enables us to write

$$\frac{T}{J} = G\frac{\partial\theta}{\partial x}. \tag{1.246}$$

Newton's second law allows us to write

$$\text{Net torque} = \left(T + \frac{\partial T}{\partial x}\delta x\right) - T = I\ddot{\theta}. \tag{1.247}$$

Therefore,

$$\frac{\partial T}{\partial x}\delta x = \rho A k^2 \delta x\,\ddot{\theta}. \tag{1.248}$$

From equation (1.246) where $J = Ak^2$ we get

$$T = Ak^2 G\theta' \tag{1.249}$$

So,

$$T' = \frac{\partial}{\partial x}(Ak^2 G\theta'). \tag{1.250}$$

Equating (1.248) and (1.250) produces

$$\frac{\partial}{\partial x}(Ak^2 G\theta') = \rho Ak^2\ddot{\theta}. \tag{1.251}$$

Thus,

$$\theta'' = \frac{\rho}{G}\ddot{\theta}. \tag{1.252}$$

We take a free vibration form of solution similar to (1.226) so that we have

$$\theta(x, t) = \phi(x) \sin \omega t. \tag{1.253}$$

Substitution of this into equation (1.252) yields the following

$$\phi'' + \lambda^2 \phi = 0, \tag{1.254}$$

where

$$\lambda = \omega \sqrt{\frac{\rho}{G}}. \tag{1.255}$$

The solution of equation (1.254) is simply

$$\phi(x) = A \cos \lambda x + B \sin \lambda x. \tag{1.256}$$

Again we have a general situation for which specific boundary conditions can be applied.

1.8.4 Boundary conditions for the shaft torsion problem

The two elementary boundary conditions for a uniform round shaft are

Free end $\qquad T = 0, \qquad \phi' = 0,$ $\qquad$ (1.257)

Clamped end $\qquad \theta = 0, \qquad \phi = 0.$ $\qquad$ (1.258)

EXAMPLE

Fixed-Free shaft:

$$x = 0, \phi = 0 \qquad \text{and} \qquad x = l, \phi' = 0. \tag{1.259}$$

Conditions (1.259) are substituted into solution (1.256) to give

$$0 = A \qquad \text{and} \qquad 0 = -\lambda A \sin \lambda x + \lambda B \cos \lambda x. \tag{1.260}$$

Therefore,

$$\lambda B \cos \lambda l = 0. \tag{1.261}$$

Since $B = 0$ then $\cos \lambda l = 0$, and for this we find that

$$\lambda l = 0, \frac{\pi}{2}, \frac{3\pi}{2}, \frac{5\pi}{2}, \ldots = \frac{n\pi}{2}, \tag{1.262}$$

where $n = 1, 3, 5, \ldots$.

Also, (1.255) gives us λ and so we obtain an expression for ω,

$$\omega = \frac{n\pi}{2l} \sqrt{\frac{G}{\rho}}. \tag{1.263}$$

The fundamental is given by $n = 1$, therefore,

$$\omega_1 = \frac{\pi}{2l} \sqrt{\frac{G}{\rho}}. \tag{1.264}$$

The associated mode shape is $\phi_1(x) = B \sin \lambda x$.

The case for which $n = 0$ is described as the rigid body mode, where motion of the system is identical at all points throughout. For the first mode $\lambda = \pi/2l$, therefore,

$$\phi(x) = B \sin\left(\frac{\pi x}{2l}\right). \tag{1.265}$$

Both equations (1.229) and (1.256) are applicable to any suitable beam and shaft problem for which the straightforward boundary conditions apply. As a rule most practical situations will reasonably closely approximate to the specific boundary conditions quoted and in this context at least the foregoing theory can be of much use.

1.8.5 The general solution

Having discussed the spatial functions $\phi(x)$ for beams and shafts in terms of two specific examples we can move on to investigate the general solutions. In all cases we regard the general solution as a function of space and time such that

$$w(x, t) = \sum_{1}^{\infty} q_n(t)\phi_n(x). \tag{1.266}$$

The $q_n(t)$ are the, as yet, unknown time dependent coordinates which must satisfy the equation

$$q_n(t) + \omega_n{}^2 q_n(t) = \frac{F_n(t)}{M_n}, \tag{1.267}$$

where $F_n(t)$ is a generalized force for the nth mode and M_n is the effective mode-dependent mass.

For the lateral vibration of beams we have the following,

$$F_n(t) = \int_0^l f(x, t)\phi_n(x)\,dx \tag{1.268}$$

and

$$M_n = \int_0^l \rho A \phi_n{}^2(x)\,dx. \tag{1.269}$$

Similarly it can be shown that the following apply for the torsional problem

$$F_n(t) = \int_0^l T(x, t)\phi_n(x)\,dx \tag{1.270}$$

and

$$M_n = \int_0^l \rho J \phi_n{}^2(x)\,dx. \tag{1.271}$$

In equations (1.268) and (1.270) $f(x, t)$ and $T(x, t)$ are, respectively, the force and torque acting, as appropriate to the problem; usually force or torque per unit length.

The solution to equation (1.267) is well known and is of the form

$$q_n(t) = A_n \cos \omega_n t + B_n \sin \omega_n t + \text{P.I.}, \qquad (1.272)$$

where P.I. denotes the particular integral; dependent of course on the form of the excitation acting (if any). The coefficients A_n and B_n may be evaluated

$$A_n = \frac{\int_0^l r(x, 0)\phi_n(x)\,dx}{\int_0^l \phi_n^2(x)\,dx}, \qquad B_n = \frac{\int_0^l s(x, 0)\phi_n(x)\,dx}{\omega_n \int_0^l \phi_n^2(x)\,dx}, \qquad (1.273)$$

where

$$r(x, 0) = y(x, 0) = \theta(x, 0)$$

and $\qquad\qquad\qquad\qquad\qquad\qquad\qquad\qquad\qquad\qquad\qquad$ (1.274)

$$s(x, 0) = y(x, 0) = \theta(x, 0)$$

(as long as the particular integral is zero at time zero).

In Section 1.6.1 the Dirac delta function was used to describe a pulse, and we can also use it here to deal with the point load or torque at some spatial position $x = e$, such that the excitation functions of equations (1.268) and (1.270) become

$$f(x, t) = f_e(t)\delta(x - e),$$
$$\qquad\qquad\qquad\qquad\qquad\qquad\qquad\qquad (1.275)$$
$$T(x, t) = T_e(t)\delta(x - e).$$

By analogy with equation (1.73) we get the following result

$$F_n(t) \quad \begin{Bmatrix} f_e(t)\phi_n(e) \\ T_e(t)\phi_n(e) \end{Bmatrix}. \qquad (1.276)$$

One assumption in determining the form of the P.I. of equation (1.272) is to regard the inhomogeneous term of equation (1.267) as harmonic such that

$$q_n(t) = \omega_n^2 q_n(t) = W \sin \Omega t, \qquad (1.277)$$

which in turn leads to the P.I. given below in the modified equation (1.272)

$$q_n(t) = A_n \cos \omega_n t + B_n \sin \omega_n t + \frac{W \sin \Omega t}{\omega_n^2 - \Omega^2}. \qquad (1.278)$$

Having established the $\phi_n(x)$, A_n, B_n, P.I., $q_n(t)$ it is then possible to construct the full general solution in the series form of equation (1.266) (provided suitable initial conditions are available at the stage of equation (1.274)). The reader is referred to the standard vibration texts, i.e.

Meirovitch (1986), Thomson (1981), Tse, Morse and Hinkle (1978) for wider treatments and examples.

It is most important that the significance of superposition (as expressed in equation (1.266)) is fully appreciated. Section 1.7.3, (equation (1.170)), showed that the general solution to the discretized problem could be obtained in much the same way. It is in many respects one of the most powerful facilities of linear vibration analysis because it allows a construction of a general solution which embraces all possible vibrational responses of the structure (for the case in question) in one expression. Thus, we additively combine all the normal modes from 1 to n to obtain an overall aggregate.

Situations in which straigthforward continuous theoretical models may become inappropriate are those where there are pronounced non-uniformities in the distribution of mass or stiffness (or both). In cases such as these we have to resort to alternative approximate approaches which combine aspects of the multi-degree of freedom techniques of lumped parameter problems (Section 1.7) with continuous theory. The ensuing methods are applicable to those problems which are not convincingly modelled by a discretized representation, and which are also too non-uniform for successful treatment by continuous methods. A tapering beam, perhaps with an end mass (where the beam mass is not itself negligible), is an example of a situation where the following analytical methods might be more appropriate. One might also consider torsional problems with significant inertial discontinuities for such treatment.

1.8.6 Methods of approximate analysis in continuous systems

(a) Rayleigh's quotient

This is a fundamental frequency estimate, applicable both to discrete and continuous models, and is developed from Rayleigh's principle which asserts that a conservative system exhibits a stationary value of oscillation in the vicinity of a natural mode.

The method assumes conservation of energy so that the total system energy is constant, therefore equation (1.16) applies here,

$$E = T + U.$$

Taking a steady oscillation at one frequency ω then at maximum amplitude we will have zero velocity and therefore

$$E = U_{max}. \tag{1.279}$$

Conversely at zero amplitude (equilibrium point) the velocity is a maximum, hence,

$$E = T_{max}. \tag{1.280}$$

Thus,

$$T_{max} = U_{max}. \tag{1.281}$$

From equation (1.18) T_{max} is defined as

$$T_{max} = \tfrac{1}{2}mC^2\omega^2 = T^*\omega^2 \tag{1.282}$$

for a single degree of freedom.

Substitution of equation (1.281) into (1.282) gives

$$\omega^2 = \frac{U_{max}}{T^*} \tag{1.283}$$

which is valid as long as T^* is suitably defined in the current context; refer to equation (1.291). If U_{max} and T^* can be estimated (we do this by using an assumed mode of vibration) then ω is in turn an estimate of the corresponding natural frequency.

For the general uniform beam case undergoing lateral vibrations the kinetic energy is given by

$$T_{max} = \tfrac{1}{2}\int_0^l \dot{y}^2 \, dm, \tag{1.284}$$

where m is the mass per unit length of beam. Thus

$$T_{max} = \tfrac{1}{2}\int_0^l \dot{y}^2 \rho A \, dx = \tfrac{1}{2}\omega^2 \int_0^l \rho A Y^2 \, dx \tag{1.285}$$

on substitution of

$$y = Y(x)\cos \omega t. \tag{1.286}$$

The potential energy function is derived by considering the work done by the elastic forces (neglecting those due to shear)

$$U = \tfrac{1}{2}\int_0^l M \, d\theta. \tag{1.287}$$

In the case of small beam deflections it can be assumed that

$$\theta = \frac{dy}{dx} \quad \text{and} \quad \frac{1}{R} = \frac{d\theta}{dx} = \frac{d^2y}{dx^2} = \frac{M}{EI}. \tag{1.288}$$

Thus

$$\frac{d\theta}{dx} = \frac{d^2y}{dx^2} \quad \text{and} \quad d\theta = y'' \, dx. \tag{1.289}$$

Therefore substitution of equations (1.288) and (1.289) into (1.287) yields

$$U = \tfrac{1}{2}EI \int_0^l (y'')^2 \, dx. \qquad (1.290)$$

We can define T^* by substituting equation (1.282) into (1.285) which gives

$$T^* = \tfrac{1}{2}\rho A \int_0^l Y^2 \, dx. \qquad (1.291)$$

U_{max} occurs when $y = Y$, therefore,

$$U_{max} = \tfrac{1}{2}EI \int_0^l (Y'')^2 \, dx. \qquad (1.292)$$

Finally, we arrive at ω from consideration of equation (1.283)

$$\omega^2 = \frac{EI \int_0^l (Y'')^2 \, dx}{\rho A \int_0^l Y^2 \, dx}. \qquad (1.293)$$

In order to progress further it is necessary to know the deflection or mode shape Y. For the fundamental frequency the static deflection curve is frequently utilized for this.

Equation (1.293) cites the general result for a straightforward example – one which should arguably be treated more effectively by either continuous or discrete theory dependent on the level of accuracy desired. Rayleigh's method offers a fundamental frequency estimation that is particularly suited to non-uniform problems such as the following tapered cantilever example in which there is a mass discontinuity in the form of a lump at the free end; this is shown in Fig. 1.25.

Since the beam tapers from the support to the free end we need to establish the flexural rigidity EI as a function of axial position x.

The beam is seen to taper to a width which is half the clamped width, therefore this function is immediately identifiable,

$$EI(x) = EI\left(1 - \frac{x}{2l}\right). \qquad (1.294)$$

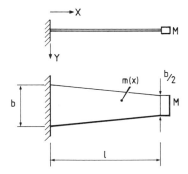

Figure 1.25 Tapered cantilevered structure with mass discontinuity in the form of a lumped mass m at l.

In a similar manner we need to specify the change in beam mass with x, again as a continuous function. This is identical in form to equation (1.294), thus,

$$m(x) = m\left(1 - \frac{x}{2l}\right), \qquad (1.295)$$

where $m(x)$ is the mass per unit length.

The beam is taken to vibrate laterally, and so coordinate y is used to express this

$$y = y(x, t).$$

Therefore we relate this to the generalized coordinate $q(t)$, by taking a likely mode shape function in the form of a simple polynomial. For the first mode we assume a quadratic, so,

$$y(x, t) = \left(\frac{x}{l}\right)^2 q(t). \qquad (1.296)$$

Equation (1.290) gives the generalized system potential energy, therefore in the present context it is

$$U = \tfrac{1}{2} \int_0^l EI(x)(y'')^2 \, dx. \qquad (1.297)$$

And so we get

$$U = \tfrac{1}{2} \int_0^l EI\left(1 - \frac{x}{2l}\right)\left\{\left[\left(\frac{x}{l}\right)^2 q(t)\right]''\right\}^2 \, dx. \qquad (1.298)$$

This becomes

$$U = 2\frac{EI}{l^4} q(t)^2 \int_0^l \left(1 - \frac{x}{2l}\right) dx = \frac{3EI}{2l^3} q(t)^2. \qquad (1.299)$$

This leads to

$$U_{max} = \frac{3EI}{2l^3} q_{max}(t)^2. \qquad (1.300)$$

The kinetic energy is a function of beam velocity and end mass velocity such that

$$T = \tfrac{1}{2} \int_0^l m(x)\dot{y}^2 \, dx + \tfrac{1}{2} M\dot{y}(l)^2. \qquad (1.301)$$

From (1.296),

$$y(l) = q(t), \qquad \dot{y}(l) = \dot{q}(t). \qquad (1.302)$$

Therefore,

$$T = \tfrac{1}{2} \int_0^l m\left(1 - \frac{x}{2l}\right)\left(\frac{x}{l}\right)^4 \dot{q}(t)^2 \, dx + \tfrac{1}{2} M\dot{q}(t)^2. \qquad (1.303)$$

Evaluating this leads to

$$T = \tfrac{1}{2}mq(t)^2\left(\frac{l}{5} - \frac{l}{12}\right) + \tfrac{1}{2}M\dot{q}(t)^2 = \tfrac{1}{2}\dot{q}(t)^2\left[\tfrac{7}{60}ml + M\right]. \quad (1.304)$$

Since $\dot{q}(t) = q_{max}\omega$ for a harmonic form of $q(t)$, we will have

$$T^* = \tfrac{1}{2}q^2{}_{max}\left[\tfrac{7}{60}ml + M\right]. \quad (1.305)$$

Equation (1.283) gives the Rayleigh estimate for ω, therefore

$$\omega^2 = \frac{U_{max}}{T^*} = \frac{3EI}{l^3\left[\tfrac{7}{60}ml + M\right]}. \quad (1.306)$$

The practical significance of the treatment of this example is considerable in that the technique allows a reasonably accurate (if slightly high) estimate of the fundamental natural frequency for the more realistic type of system – one might consider blade vibrations in particular as appropriate to this category. In addition to the usefulness of the result it is not in itself a difficult method to apply as long as the non-uniformities are mathematically expressible without too much additional effort. The question of accuracy is probably best addressed by considering a simple example which can be treated by the exact theory of Section 1.8.1 so that there is a basis for comparison.

The general beam problem discussed in Section 1.8.6 can be re-examined by assuming the static deflection curve for the fundamental mode shape such that,

$$Y(x) = \tfrac{1}{2}Y_0\left[3\left(\frac{x}{l}\right)^2 - \left(\frac{x}{l}\right)^3\right]. \quad (1.307)$$

Substituting equation (1.307) into the numerator of equation (1.293) leads to

$$EI\int_0^l (Y'')^2\,dx = \frac{3Y_0{}^2}{l^3}. \quad (1.308)$$

The denominator therefore becomes

$$\rho A\int_0^l Y^2\,dx = \frac{33Y_0{}^2}{140}\,l\rho A. \quad (1.309)$$

Finally, we have

$$\omega^2 = 12.727\,\frac{EI}{\rho A l^4}. \quad (1.310)$$

The approximate natural frequency is therefore

$$\omega = 3.567\,\sqrt{\frac{EI}{\rho A l^4}}. \quad (1.311)$$

Putting the appropriate boundary conditions ($\phi = \phi' = 0$ at $x = 0$, $\phi'' = \phi''' $ at $x = l$) into equation (1.229) leads to the exact solution for ω, namely,

$$\omega_{\text{exact}} = 3.516 \sqrt{\frac{EI}{\rho A l^4}}. \tag{1.312}$$

The Rayleigh quotient can be seen to give a result which is very close to the exact case ($\sim$1.5% high) and it is doubtful in practical terms that there is much to be gained in utilizing it in a simple example such as this. However equation (1.311) does show how accurate this approximation can be, given a suitable function for $Y(x)$.

(b) The Rayleigh–Ritz approach for natural frequencies of higher modes

The Rayleigh energy method (quotient) is highly sensitive to the accuracy of the mode shape function that is chosen. In addition to this it always generates a high estimate of the fundamental (more formally it gives an upper bound for the fundamental eigenvalue) and cannot deal with higher modes and natural frequencies. The Rayleigh–Ritz approach enables the natural frequencies and the shapes of n modes to be evaluated by the linear combination of an assumed set of deflection shapes $\psi_1(x)$, $\psi_2(x)$, ..., $\psi_n(x)$. If we restrict the analysis to the fundamental and the second mode we find that the general solution is of the form

$$w(x, t) = \psi_1(x)q_1(t) + \psi_2(x)q_2(t). \tag{1.313}$$

For the purposes of comparison the example of the uniform cantilever beam will be taken again here; it is therefore assumed that the two trial deflection shapes ψ_1 and ψ_2 are of the form

$$\psi_r(x) = \left(\frac{x}{l}\right)^{r+1}, \qquad r = 1, 2, \ldots, n. \tag{1.314}$$

The boundary conditions are $w = w' = 0$ at $x = 0$ and $M = 0$ at $x = l$, the former being relevant to this problem. Substituting equation (1.314) into (1.313) yields

$$w(x, t) = \left(\frac{x}{l}\right)^2 q_1(t) + \left(\frac{x}{l}\right)^3 q_2(t). \tag{1.315}$$

Potential energy is given by equation (1.290) requiring the second space derivative, therefore,

$$w''(x, t) = \left(\frac{2}{l^2}\right)q_1(t) + \left(\frac{6x}{l^3}\right)q_2(t). \tag{1.316}$$

Substituting this into equation (1.290) and differentiating the resulting expression respectively with respect to coordinates q_1 and q_2 produces

$$\frac{\partial U}{\partial q_1} = 2\,\frac{EI}{l^3}\,[2q_1 + 3q_2] \tag{1.317}$$

and

$$\frac{\partial U}{\partial q_2} = \frac{6EI}{l^3}\,[q_1 + 2q_2]. \tag{1.318}$$

These steps are necessary to establish the governing equations (in matrix form) by means of the Lagrangian approach; so in order to complete this we differentiate equation (1.315) once with respect to time and substitute the result in equation (1.285) (noting the notational change from $\dot{y}$ to $\dot{w}$), which leads to

$$T = \frac{\rho A}{2}\left\{\frac{l}{5}\,\dot{q}_1^{\,2} + \frac{l}{3}\,\dot{q}_1\dot{q}_2 + \frac{l}{7}\,\dot{q}_2^{\,2}\right\}, \tag{1.319}$$

where $T = T_{\max}$. Therefore

$$\frac{\mathrm{d}}{\mathrm{d}t}\left(\frac{\partial T}{\partial \dot{q}_1}\right) = \rho Al\left\{\frac{\ddot{q}_1}{5} + \frac{\ddot{q}_2}{6}\right\},$$

$$\frac{\mathrm{d}}{\mathrm{d}t}\left(\frac{\partial T}{\partial \dot{q}_2}\right) = \rho Al\left\{\frac{\ddot{q}_1}{6} + \frac{\ddot{q}_2}{7}\right\}, \tag{1.320}$$

since $\partial T/\partial q_{1,2} = 0$ the equations of motion are

$$\rho Al\begin{bmatrix} 1/5 & 1/6 \\ 1/6 & 1/7 \end{bmatrix}\ddot{\bar{q}} + \frac{EI}{l^3}\begin{bmatrix} 4 & 6 \\ 6 & 12 \end{bmatrix}\bar{q} = 0. \tag{1.321}$$

The eigenvalue problem now emerges if we state that

$$\bar{q} = \bar{X}\mathrm{e}^{i\omega t}. \tag{1.322}$$

Substitution of equation (1.322) into (1.321) gives

$$\left\{\begin{bmatrix} 4 & 6 \\ 6 & 12 \end{bmatrix} - \frac{\omega^2 \rho Al^4}{EI}\begin{bmatrix} 1/5 & 1/6 \\ 1/6 & 1/7 \end{bmatrix}\right\}\bar{X} = 0. \tag{1.323}$$

Evaluating and setting the characteristic determinant to zero provides the required frequency equation

$$\left(\frac{\omega^2 \rho Al^4}{EI}\right)^2 - 1224\,\frac{\omega^2 \rho Al^4}{EI} + 15\,120 = 0. \tag{1.324}$$

From this it is easily established that

$$\omega_1 = 3.53\,\sqrt{\frac{EI}{\rho Al^4}}\,, \qquad \omega_2 = 34.8\,\sqrt{\frac{EI}{\rho Al^4}}. \tag{1.325}$$

Equation (1.312) may be directly compared with ω_1 in (1.325) and equation (1.311) (Rayleigh quotient) with the result that the Rayleigh–Ritz estimate lies between the exact case and the Rayleigh quotient value. By substituting the required boundary conditions into equation (1.229) (i.e. those used to obtain equation (1.312)), and setting $n = 2$ we get the exact result for the second mode

$$\omega_{2\text{exact}} = 22.0 \sqrt{\frac{EI}{\rho A l^4}}. \tag{1.326}$$

It is quite clear that the Rayleigh–Ritz method is very much less accurate for the second mode than the fundamental. This is wholly a function of the form of equation (1.315) which in this example gives a result which is around 58% too high!

Further manipulation of equation (1.323) leads to

$$\begin{bmatrix} (4 - \eta/5) & (6 - \eta/6) \\ (6 - \eta/6) & (12 - \eta/7) \end{bmatrix} \bar{X} = 0, \tag{1.327}$$

where

$$\eta = \frac{\Omega^2 \rho A l^4}{EI} \tag{1.328}$$

so that we get,

$$\frac{X_2}{X_1} = -\left(\frac{4 - \eta/5}{6 - \eta/6}\right). \tag{1.329}$$

For the fundamental normal mode we have $\eta_1 = 12.46$ which yields

$$\begin{bmatrix} X_1 \\ X_2 \end{bmatrix}_1 = \begin{bmatrix} 1 \\ -0.38 \end{bmatrix} \tag{1.330}$$

and for the second mode, $\eta_2 = 1211.0$, and therefore

$$\begin{bmatrix} X_1 \\ X_2 \end{bmatrix}_2 = \begin{bmatrix} -0.82 \\ 1 \end{bmatrix}. \tag{1.331}$$

Because of the form of solution (1.322) it is appropriate to insert the eigenvectors X in place of the $q_i(t)$ in equation (1.315). This produces a general solution of the form

$$w_1(x, t) = \left(\frac{x}{l}\right)^2 - 0.38\left(\frac{x}{l}\right)^3 \tag{1.332}$$

for the fundamental, and similarly for the second mode,

$$w_2(x, t) = -0.82\left(\frac{x}{l}\right)^2 + \left(\frac{x}{l}\right)^3. \tag{1.333}$$

These are shown in Fig. 1.26.

There is clearly no node ($w_1 = 0$) for the fundamental (given by

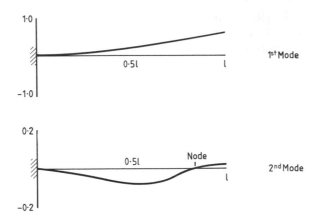

Figure 1.26 First and second mode shapes for Rayleigh–Ritz model of a uniform cantilever beam.

equation (1.332)) in the range $0 < x < l$ but for the second mode a node is evident at $x = 0.82l$. This is verified by Fig. 1.26.

In essence the method works by the superpositioning of two trial shape estimates from which a form of the free vibration eigenvalue problem is derived leading to the requisite natural frequencies and mode shapes. The example quoted shows that the simple polynomial estimates give a very good approximation to the fundamental, however the accuracy of approximation to the second natural frequency and mode is low. This emphasizes the sensitivity of the method to the trial shape estimates.

We can develop the Rayleigh–Ritz technique further so that any number of frequencies and modes can be calculated; deriving $w(x, t)$ from the vector dot product thus

$$w(x, t) = \bar{\psi}(x)\bar{q}(t), \tag{1.334}$$

where

$$\bar{\psi} = [\psi_1, \ldots, \psi_n] \quad \text{and} \quad \bar{q} = \begin{bmatrix} q_1 \\ \vdots \\ q_n \end{bmatrix}.$$

Therefore,

$$w'' = \bar{\psi}''\bar{q} \tag{1.335}$$

and so

$$U = \tfrac{1}{2}EI \int_0^l (\bar{\psi}''\bar{q})^2 \, \mathrm{d}x. \tag{1.336}$$

As $\bar{\psi}''\bar{q}$ is a scalar we can write the following

$$(\bar{\psi}''\bar{q})^2 = \bar{q}^{\mathrm{T}}\bar{\psi}''^{\mathrm{T}}\bar{\psi}''\bar{q}. \tag{1.337}$$

Also q is not a function of x, therefore,

$$U = \tfrac{1}{2}EI\bar{q}^{\mathrm{T}}\left(\int_0^l \bar{\psi}''^{\mathrm{T}}\bar{\psi}'' \,\mathrm{d}x\right)\bar{q}. \tag{1.338}$$

By differentiating U with respect to $\bar{q}$ we can show that the generalized stiffness $\bar{K}$ (where $\partial U/\partial\bar{q} = \bar{K}\bar{q}$) is given by

$$\bar{K} = EI\int_0^l \bar{\psi}''^{\mathrm{T}}\bar{\psi}'' \,\mathrm{d}x. \tag{1.339}$$

Taking the uniform beam case once more

$$T = \tfrac{1}{2}\rho A \int_0^l (\bar{\psi}\bar{q})^2 \,\mathrm{d}x \tag{1.340}$$

leading to

$$\frac{\mathrm{d}}{\mathrm{d}t}\left(\frac{\partial T}{\partial\bar{q}}\right) = \rho A\left(\int_0^l \psi^{\mathrm{T}}\psi \,\mathrm{d}x\right)\bar{\bar{q}}. \tag{1.341}$$

Therefore the generalized mass is

$$\bar{M} = \rho A\int_0^l \bar{\psi}^{\mathrm{T}}\bar{\psi} \,\mathrm{d}x. \tag{1.342}$$

To use these generalized results for the uniform beam problem we take a trial $\bar{\psi}$

$$\bar{\psi} = \left[\left(\frac{x}{l}\right)^2 \left(\frac{x}{l}\right)^3 \left(\frac{x}{l}\right)^4\right]. \tag{1.343}$$

Therefore,

$$\bar{\psi}''^{\mathrm{T}}\bar{\psi}'' = \left(\frac{2}{l^2}\right)^2 \begin{bmatrix} 1 \\ 3x/l \\ 6x^2/l^2 \end{bmatrix}\left[1 \quad \left(\frac{3x}{l}\right) \quad \left(\frac{6x^2}{l^2}\right)\right] \tag{1.344}$$

from which we get

$$\bar{K} = \frac{EI}{l^3}\begin{bmatrix} 4 & 6 & 8 \\ 6 & 12 & 18 \\ 8 & 18 & 28.8 \end{bmatrix}. \tag{1.345}$$

Substitution of equation (1.343) into (1.344) produces the generalized mass matrix for this problem

$$\bar{M} = \rho Al\begin{bmatrix} 1/5 & 1/6 & 1/7 \\ 1/6 & 1/7 & 1/8 \\ 1/7 & 1/8 & 1/9 \end{bmatrix}. \tag{1.346}$$

Now the equation analogous to (1.321) is constructed and solved, thus,

$$\bar{M}\bar{\bar{q}} + \bar{K}\bar{q} = 0. \tag{1.347}$$

This can be either by iteration or by computer program and will clearly result in a third-order polynomial for the frequency equation. The second normal mode frequency, upon evaluation, turns out to be

$$\omega_2 = 21.4 \sqrt{\frac{EI}{\rho A l^4}},\tag{1.348}$$

which is very much nearer the exact answer of equation (1.326) than the previous estimate.

In conclusion it can be seen that the inclusion of ψ_3 in the trial shape function provides a much more accurate convergence of the frequency equation for the second natural frequency, thus it would seem sensible to (at least) include ψ_{n+1} in the computation for all ω_n when using this method.

2

Parametric vibrations in linear vibrating systems

INTRODUCTION

This chapter is devoted to a discussion of parametric vibrations in mechanical systems, and examines the nature of stability in parametric problems that are linear, or those for which linear approximations are acceptable. The term **parametric** is descriptive of cases where for some reason the external excitation action appears as a time varying modification of a system parameter. This is in marked contrast to the forced vibration problem where energy is simply fed into the system and the system responds, or not, dependent on the resonance condition (if any) in operation. Forced vibration problems do not show parameter variation as a result of imposed forcing. On the other hand a parametrically excited system will show variation of a parameter (such as restoring force, for example) with the excitation as long as an appropriate resonance is in force. This proviso is all important; a non-resonant parametrically excited system will not normally respond, except perhaps as a decaying transient dependent upon initial conditions.

Some typical 'parametric' configurations are given in Fig. 2.1; these are somewhat simplified engineering structures and systems, for which certain satisfied (or near satisfied) relationships between the natural frequencies and the frequency of excitation can be arranged, and for which there is a low inherent damping, and also a parameter varying harmonic excitation. As a first example we will go on to examine a simple cantilever structure under parametric base excitation. This will allow a clear picture of important phenomena to be created.

In general mathematical terms the inhomogeneous differential equations of motion for a forced system are replaced by homogeneous forms,

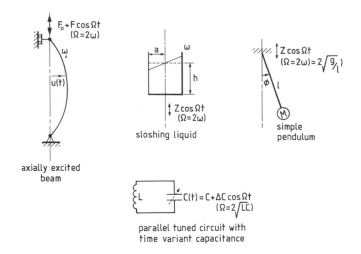

axially excited
beam

sloshing liquid

simple
pendulum

parallel tuned circuit with
time variant capacitance

Figure 2.1 Examples of physical systems exhibiting potential parametric resonance effects (after Ibrahim and Barr, 1978).

in which there exist varying, periodic, coefficients, in the parametric case. This is typified by the fundamental Mathieu–Hill equation

$$\ddot{q} + (a + b \cos \Omega t)q = 0. \tag{2.1}$$

A more generalized version for which the excitation function $f(t)$ is periodic, but not necessarily harmonic, is the Hill equation,

$$\ddot{q} + (a + f(t))q = 0. \tag{2.2}$$

Parametric systems, as already stated, respond when the frequency of excitation is related to the natural frequency (or frequencies) by a resonance condition, and this does not imply synchronicity between these frequencies. So, large responses may be generated in cases where the excitation frequency is remote (but related through an integer – or fractional – multiple of some sort) from the system natural frequency, or frequencies. The precise nature of such frequency relationships will form the subject of further discussion.

The motion that results from a parametric excitation is unstable and grows, exponentially, with time. The eventual magnitude of the response is not directly governed by the damping acting within the system, but by the effects of extreme displacements such as nonlinear (cubic) stiffness. Nonlinear stiffness tends to predominate over the usual linear form in many structural and machine cases when the displacement magnitude is large enough. Where the damping often appears to be significant in parametric problems is in the early stages of response build up; a lower level of damping will lead to an early and vigorous response

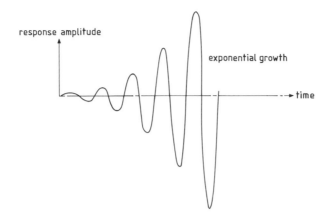

Figure 2.2 Parametric response growth with time for a cantilevered beam.

at a lower level of excitation. Conversely a higher level of damping will inhibit the speed of response build up and will raise the excitation threshold necessary to generate a response, however it will not, generally, affect the final response amplitude. There have even been reports of cases in which the damping in certain parametrically resonant systems has assisted the response motion. The behaviour of parametric response with time for perhaps a simple cantilever beam undergoing support motion is shown in Fig. 2.2.

The foregoing deals exclusively with linear parametric problems; those that are typified by forms of equation (2.1), and it is fundamental to understanding the relevance of linearity in these instances that we appreciate that systems will respond, theoretically, in a completely unbounded manner. Physical, limiting, nonlinear effect almost always come into play in practice, however, and are mainly responsible for the apparently steady motion of a parametrically driven component or structure. Thus an 'unstable' system will in actuality exhibit a limit cycle, and so it settles down into steady state response.

2.1 STABILITY CONSIDERATIONS

The stability of the Mathieu–Hill class of equations was explored towards the end of the last century (Floquet, 1883), but is still central to our appreciation of linear parametric problems, and so a shortened form of the analysis will be given here. If we take the general Mathieu–Hill equation (2.1) then is it clear from linear systems theory that there can be two linearly independent solutions to this equation, $f_1(t)$ and $f_2(t)$, which can be superpositioned,

$$f(t) = Af_1(t) + Bf_2(t) \tag{2.3}$$

A and B are constants; $f(t)$ is the complete solution.

From equation (2.1) the period of excitation is obviously

$$T = 2\pi/\Omega \tag{2.4}$$

and solutions $f_1(t)$ and $f_2(t)$ could be time shifted to become $f_1(t + T)$, $f_2(t + T)$. Further to this we could regard the time shifted solutions to be linear combinations of the original solutions, thus,

$$f_1(t + T) = a_{11}f_1(t) + a_{12}f_2(t), \tag{2.5}$$

$$f_2(t + T) = a_{21}f_1(t) + a_{22}f_2(t). \tag{2.6}$$

Returning to the original solutions $f_1(t)$ and $f_2(t)$ we could define these as linear combinations of some other, unspecified, solutions $f_1*(t)$, and $f_2*(t)$. Therefore

$$f_1(t) = c_{11}f_1*(t) + c_{12}f_2*(t),$$
$$f_2(t) = c_{21}f_1*(t) + c_{22}f_2*(t). \tag{2.7}$$

After some further manipulation we arrive at

$$\begin{bmatrix} f_1*(t + T) \\ f_2*(t + T) \end{bmatrix} = [c]^{-1}[a][c] \begin{bmatrix} f_1*(t) \\ f_2*(t) \end{bmatrix}. \tag{2.8}$$

By making a judicious choice for the form of matrix $[c]$ it is quite possible to arrange for $[c]^{-1}[a][c]$ to be a diagonal matrix which we could designate as $[R]$, so that,

$$[R] = \begin{bmatrix} \eta_1 & 0 \\ 0 & \eta_2 \end{bmatrix}. \tag{2.9}$$

Thus η_1 and η_2 are the eigenvalues of $[a]$. Substitution of equation (2.9) into (2.8) leads to

$$f_1*(t) = \eta_1 f_1*(t + T),$$
$$f_2*(t) = \eta_2 f_2*(t + T), \tag{2.10}$$

and the characteristic equation

$$\begin{vmatrix} (a_{11} - \eta) & a_{12} \\ a_{21} & (a_{22} - \eta) \end{vmatrix} = 0. \tag{2.11}$$

The original independent solutions $f_{1,2}(t)$ are unknown, but we can assume that they can be made to satisfy linearly independent initial conditions, such as,

$$f_1(0) = \dot{f}_2(0) = 1,$$
$$\dot{f}_1(0) = f_2(0) = 0. \tag{2.12}$$

Equations (2.12) may be substituted into equations (2.6) and (2.7) (differentiating where required), to give

$$a_{11} = f_1(T); \quad a_{21} = f_2(T); \quad a_{12} = \dot{f}_1(T); \quad a_{22} = \dot{f}_2(T). \quad (2.13)$$

Therefore the characteristic equation (2.11) becomes,

$$\eta^2 - 2\alpha\eta + \beta = 0, \quad (2.14)$$

where

$$\alpha = \tfrac{1}{2}(f_1(T) + \dot{f}_2(T)),$$
$$\beta = f_1(T)\dot{f}_2(T) - \dot{f}_1(T)f_2(T), \quad (2.15)$$

β can be shown to be equal to unity if we substitute

$$q = \left\{ \begin{matrix} f_1(t) \\ f_2(t) \end{matrix} \right.$$

in equation (2.1) and use the initial conditions of (2.12) in the ensuing analysis. Thus equation (2.14), where $\beta = 1$, is the characteristic equation for two linearly independent solutions to Bolotin's form of the Mathieu–Hill type of equation. The roots can be evaluated such that

$$\eta_1 = \alpha + \sqrt{\alpha^2 - 1}; \quad \eta_2 = \alpha - \sqrt{\alpha^2 - 1}. \quad (2.16)$$

Closer examination of forms (2.16) shows that

$$\eta_1 \eta_2 = 1 \quad (2.17)$$

for which case one of the following must apply.

1. The roots are real, of the same sign, but with different magnitudes such that one has an absolute value > 1 and the other < 1.
2. The roots are real, and are $\eta_1 = \eta_2 = \pm 1$.
3. The roots are a pair of complex conjugates with unity modulus.

These stability cases can be explored by examining the loci of the roots in the complex plane. This is shown in Fig. 2.3. The general physical interpretations of cases (1)–(3) are therefore:

1. The system is unstable and will respond, one response growing exponentially with time, the other declining to zero.
2. The system is on the threshold of stability and instability, its behaviour with time is dependent on external conditions.
3. The system is stable with no unboundedness of response apparent, despite the possible presence of large external perturbations.

It is usual to express the stability of these systems in terms of useful physical parameters such as excitation magnitude and excitation frequency. In this way we map the system on to a two parameter plane

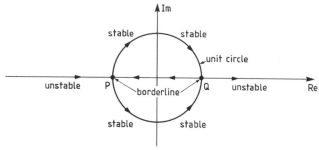

(i) both η_1, η_2 are real; $\eta_1 > 1, \eta_2 < 1$ or vice versa (along Re axis)

(ii) both η_1, η_2 are real; $\eta_1 = \eta_2 = 1$ (points P and Q)

(iii) both η_1, η_2 are complex; unity modulus (somewhere on unit circle)

Figure 2.3 Locus of the roots of the characteristic equation (2.14) defining solution stability in the complex plane.

which, importantly, has direct physical significance to the problem. The plane is shown as the stability chart or sometimes as the Strutt–Ince diagram, and an example depicting resonance around the principal parametric zone, for which,

$$\Omega = 2\omega \qquad (2.18)$$

is given in Fig. 2.4. This is for one particular version of the Mathieu–Hill equation as shown on the figure.

The important feature of the stability chart is the transition curve, a curve which delineates the chart into areas of stability or instability. Therefore a pair of $(\mu, \Omega/2\omega)$ values, which define a coordinate within the shaded region, define that the system for which they have been

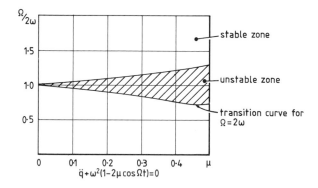

Figure 2.4 Stability chart showing the zoning for principal parametric resonance of the Mathieu–Hill equation.

calculated will be unstable. On the other hand, a $(\mu, \Omega/2\omega)$ coordinate outside the shaded area defines a stable response. This stable response will depend on the initial conditions and on the presence (if any) of external excitation in the form of a right-hand side forcing term. In the absence of such an additional excitation the stable response will decay to zero from the initial conditions. Points on the transition curve define borderline stability. Further scrutiny of Fig. 2.4 reveals that the transition curve broadens out with increasing μ. Thus unstable, non-zero, responses are predicted for values of Ω which do not exactly satisfy equation (2.18). Therefore the higher the excitation amplitude the greater the permissible detuning (region around Ω) for which instability might be expected. It is common practice to express equation (2.18) in a revised form to incorporate this modification

$$\Omega = 2\omega + \varepsilon\sigma, \qquad (2.19)$$

where $\varepsilon\sigma$ is some degree of detuning around 'perfect' principal parametric resonance, and is usually loosely considered to be small.

The techniques for deriving transition curves for parametric problems all depend on identifying an appropriate analysis with which, ideally, an expression for the detuning parameter as a function of system constants (i.e. damping ratios, natural frequencies) may be found. As the specific non-zero solutions to linear parametric problems are unbounded with time in the unstable zone specialized methods are necessary, and considerable effort has already been expended by several prominent researchers in the elucidation of these. Nayfeh and Mook (1979) pinpointed three distinct categories of suitable analytical methods and rigorously listed their principal proponents. The first method is attributable to Hill (1886) and is known as the method of infinite determinants. Application of this technique to parametric vibration problems necessitates a certain amount of approximation in dealing with the infinite set of linear, algebraic, homogeneous equations that results from taking a Fourier series solution to the governing differential equation. Clearly then the coefficient matrix, which we can derive from this set, will be of infinitely large dimensions, as will be its determinant (which must equate to zero for non-zero solutions). Fortunately the method will give reasonable results if the elements of the most central rows and columns are considered, and the rest neglected. This approximation leads to relative computational ease of application, but experience has shown that there is no clear indication of the relationship between the chosen order of the reduced determinant and that of the resulting transition curve expansion. The difficulty with this is that it will lead to sizeable errors in second- and higher-order computations, and so will clearly tend to invalidate efforts to investigate anything other that the 'strongest' first-order resonant effects.

The consequence of the ordering of terms in differential governing equations for parametric systems will be discussed further, refer to Section 2.2.

The next method of stability assessment is due to Liapunov (Lasalle and Lefschetz, 1961; Jordan and Smith, 1979), and in the context of parametric problems could be used to give a qualitative indication of system stability. The method is based upon the premise that system energy does not increase near an equilibrium state and as such Liapunov's ideas can be seen as extensions to the general energy concept. One applies the method by examining the properties of Liapunov functions from a knowledge of the whole system rather than the solutions. The rules for finding suitable Liapunov functions are complex, but can be applied fairly straightforwardly to certain, relatively simple, practical problems. Because the results are essentially qualitative Liapunov's direct method will not be examined in further detail in this chapter.

Finally we come to the more directly usable analytical methods such as Struble's asymptotic approximations method, averaging methods (including the method of harmonic balance), the Linstedt–Poincaré and also the multiple scales perturbation techniques. Struble's method is really a perturbational procedure and therefore belongs, generically, with Lindstedt–Poincaré etc. These methods, and others are all potentially capable of successfully treating parametric and nonlinear problems, however there may be certain drawbacks in particular individual applications. For example the techniques under the general heading of averaging (including harmonic balance, and the Krylov–Bogoliubov, and Krylov–Bogoliubov–Mitropolski methods) all require some advance expectations of the form of the result. In addition to this problem plausible solutions could be generated, which, on checking with a more dependable method, could be shown to be substantially inaccurate. Therefore all averaging methods are, at least, potentially inconsistent and should be used with caution. On the other hand these methods can be relatively straightforward and generally require rather less analytical effort than their perturbational counterparts; this is a distinct advantage in their favour. The pertubation techniques are somewhat cumbersome algebraically and the manipulative effort needed can quickly become prohibitive as system complexity and accuracy requirements begin to increase. Their strengths, however, lie in the general dependability of the results obtained, which (whilst there is always a clear degree of approximation involved) are usually consistent with the level of accuracy looked for. The various analytical tools available for parametric and nonlinear engineering problems have been closely documented and explained in the literature (Nayfeh, 1973; Jordan and Smith, 1979; Nayfeh and Mook, 1979; Schmidt and Tondl, 1986). The references given are detailed and definitive in their coverage of the mathematics of most of the well-known methods.

The transition curve for the principal parametric resonance case of Fig. 2.4 can be obtained using either an averaging or a pertubational approach to derive an expression for $\varepsilon\sigma$ in equation (2.19). In Section 2.3 we use a further example to illustrate the methods of harmonic balance and multiple scales, and for this a cantilevered structure undergoing base excitation is considered in such a way that combination resonances are generated.

2.2 SOME SIMPLE PHYSICAL SYSTEMS

Because there is a degree of mathematical complexity inherent in all parametric problems it is important to examine fundamental physical systems in advance of more difficult problems – and to this end a simple structural cantilever model is initially proposed, as depicted in Fig. 2.5 in two orientations. Figure 2.5(a) shows the beam arranged so that the sinusoidal excitation is applied axially. In practice, it is convenient to

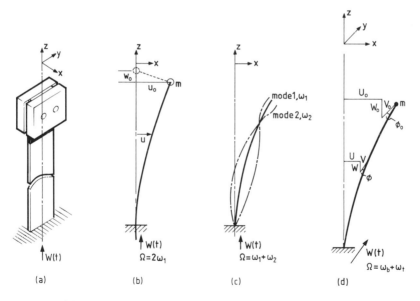

Figure 2.5 (a) Cantilever model with end mass and coordinate reference frame; (b) shows schematic with displacements and excitation function for principal parametric resonance ($\Omega = 2\omega_1$); (c) shows schematic of combinatory response for bending/bending combination resonance $\Omega = \omega_1 + \omega_2$; (d) shows kinematical coupling provided by v_0 displacement for bending/torsion combination resonance $\Omega = \omega_b + \omega_t$.

include a lumped mass at the free end of the beam to facilitate tuning and to ensure a practical range between the lower natural frequencies. The lumped mass also means that we can model the problem as one with one degree of freedom (or maybe two for combinatory motion involving torsion). Thus we will neglect the mass of the beam as compared with the lumped mass, and will consider the beam purely as a flexural element. This simple modelling approach enables the problem to be kept to its most fundamental so that the phenomena of interest are not lost in the mathematics. The deformed model is shown in Fig. 2.5(b). Lateral motion of the beam may be in terms of the fundamental bending mode (as shown) or possibly as a combination of the fundamental and higher bending modes as shown in Fig. 2.5(c). Physically the successful excitation of a combination resonance depends upon the excitation frequency being close to the sum or difference of the combinatory (natural) frequencies. For example

$$\Omega = (m\omega_a \pm n\omega_b) \pm \varepsilon\rho, \qquad (2.20)$$

where m and n are integers.

It also depends on the excitation magnitude and the system damping achieving appropriate levels in order to initiate the instability. We could examine combinatory motion in practice by observing the behaviour of a simple laboratory model under a strobe light with an adjustable repetition rate. One is not limited to combinatory pairs only, and three and more frequencies may be seen to participate (Cartmell and Roberts, 1987). Whether or not these multifrequency combinations are excited depends entirely on the system parameters. As a rule combinations involving more than two frequencies are generally more elusive and need relatively higher levels of excitation to initiate the instability. Because of this they could be regarded as less important than the single frequency effects (principal parametric resonance) or the 'strong' two frequency combinations, however this is not so because the corresponding instability, once generated, will be just as vigorous (and potentially destructive) as that of an easier excited effect. Some recognition of the relative significance of terms in the governing equations of motion is needed and to do this we order the terms by means of some small (sometimes arbitrary) ordering parameter. Conventionally ε is used and is introduced in such a way that it appears as a coefficient of the parametric excitation term(s) and also the damping – if this is also to be considered. Ordered terms may be linear in ε, or to higher order depending on how the terms in question are regarded. It is difficult to be precise about the choice of ordering when constructing equations, however a general rule is that the 'weaker' terms are considered to be of higher order. The examples which follow should help to clarify this to some extent.

2.2.1 System under principal parametric resonance

If we consider the example of Fig. 2.5(b) it is clear that the total kinetic energy will be given by

$$T = \tfrac{1}{2}m_0(\dot{u}_0{}^2 + [\dot{w}_0 - \dot{W}]^2),$$ (2.21)

where u_0, w_0, W are, respectively, the lateral displacement of the end mass, the axial contraction displacement of the mass centre and the excitational displacement at the support. Note that there is no kinetic energy contribution due to rotary inertia since we are only considering a point mass. The potential energy is due principally to strain energy in bending and is given by the following (Timoshenko *et al.*, 1974)

$$U = \tfrac{1}{2}EI_y \int_0^l (u'')^2 \, dz.$$ (2.22)

We can relate the general lateral displacement u to that at the mass centre, u_0, by using the following relationship based on the linear shape function $f(z)$, for bending. So

$$u(z, t) = f(z)u_0(z, t).$$ (2.23)

Substitution of equation (2.23) into (2.22) leads to

$$U = \tfrac{1}{2}EI_y \left[\int_0^l (f'')^2 dz \right] u_0{}^2.$$ (2.24)

Thus we have the strain energy in terms of the lateral displacement at the mass centre u_0.

The contraction displacement is directly related to the lateral displacement (Timoshenko, Young and Weaver, 1974)

$$w = \tfrac{1}{2} \int_0^l (u')^2 \, dz.$$ (2.25)

After substituting equation (2.23) in (2.25) we get

$$w_0 = \tfrac{1}{2} \left[\int_0^l (f')^2 \, dz \right] u_0{}^2.$$ (2.26)

It should be mentioned that a gravitational potential energy term could be added to equations (2.22) and (2.24) to complete the model more fully. In this case we would have,

$$U = \tfrac{1}{2}EI_y \left[\int_0^l (f'')^2 \, dz \right] u_0{}^2 - m_0 g w_0.$$ (2.27)

This would effectively change the computed natural frequency, however the modification would be minor and we will not develop it further here. Differentiating equation (2.26) with respect to time leads to

$$\dot{w}_0 = \left[\int_0^l (f')^2 \, dz \right] (u_0 \dot{u}_0) = B u_0 \dot{u}_0.$$ (2.28)

This can be substituted into equation (2.21) to yield

$$T = \tfrac{1}{2}m_0\dot{u}_0^2 + \tfrac{1}{2}m_0\{B^2u_0^2\dot{u}_0^2 + \dot{W}^2 - 2Bu_0\dot{u}_0\dot{W}\}. \qquad (2.29)$$

By applying Lagrange's equations we get

$$\frac{\partial T}{\partial \dot{u}_0} = m_0\dot{u}_0 + \tfrac{1}{2}m_0\{B^2u_0^2 2\dot{u}_0 - 2Bu_0\dot{W}\} \qquad (2.30)$$

and

$$\frac{\mathrm{d}}{\mathrm{d}t}\left(\frac{\partial T}{\partial \dot{u}_0}\right) = m_0\ddot{u}_0 + \tfrac{1}{2}m_0\{B^2 2(2\dot{u}_0^2 u_0 + u_0^2\ddot{u}_0) - 2B(\dot{u}_0\dot{W} + u_0\dot{Y})\}. \qquad (2.31)$$

Also

$$\frac{\partial T}{\partial u_0} = \tfrac{1}{2}m_0\{B^2 2u_0\dot{u}_0^2 - 2B\dot{u}_0\dot{W}\}. \qquad (2.32)$$

Subtracting equation (2.32) from (2.31) leads to

$$\frac{\mathrm{d}}{\mathrm{d}t}\left(\frac{\partial T}{\partial \dot{u}_0}\right) - \frac{\partial T}{\partial u_0} = m_0\ddot{u}_0 + m_0 B^2 u_0\dot{u}_0^2 + m_0 B^2 u_0^2\ddot{u}_0 - m_0 u_0\dot{W}B. \qquad (2.33)$$

Differentiating equation (2.27) (neglecting the gravity term), with respect to u_0 leads to

$$\frac{\partial U}{\partial u_0} = EI_y\left[\int_0^l (f'')^2 \,\mathrm{d}z\right]u_0. \qquad (2.34)$$

Equations (2.33) and (2.34) can be added and equated to zero

$$m_0\ddot{u}_0(1 + B^2u_0^2) + m_0 B^2 u_0\dot{u}_0^2 + EI_y\left[\int_0^l (f'')^2 \,\mathrm{d}z\right]u_0 - m_0 u_0\dot{W}B = 0. \qquad (2.35)$$

This is the derived governing equation for the system proposed. It is clearly evident that there are two nonlinear terms (second and third), both of which are cubic in form. We will neglect these for the sake of this analysis but will return to examine the effects of terms such as these later in the book (Section 3.1.1).

So dividing through by m_0, neglecting the nonlinear terms, and adding in a viscous damping term leads to the final form of the governing equation for the parametrically excited beam configured as shown in Fig. 2.5(b)

$$\ddot{u}_0 + 2\xi\omega\dot{u}_0 + \omega^2 u_0 - B\ddot{W}u_0 = 0. \qquad (2.36)$$

This is a form of the Mathieu–Hill equation (with damping) whereby the excitation (acceleration, $\ddot{W}(t) = (W_0\Omega^2)\cos\Omega t$) acts as a coefficient of u_0. Refer to Section (2.3.4) for the resulting stability zoning.

2.2.2 Combination resonances involving bending and torsional motion

The configuration of Fig. 2.5(d) models the case where the beam bends laterally and also twists about the z axis due to a parametric excitation in the form of support motion in the stiffest plane of the beam. The response motion is a combinatory vibration in the fundamental bending and torsional modes and so this is a rather more complex problem than its predecessor. The coupling may be traced directly to the presence of v_0, a small but important displacement in the plane of the excitation, as shown in Fig. 2.5(d). By consideration of the geometry of the problem it is possible to define v_0 in terms of the model displacement coordinates u_0 and ϕ_0 by means of constraint equations so that we have

$$v_0 = B_1\phi_0 u_0,\qquad(2.37)$$

where

$$B_1 = \int_0^l (l - z)h(f'')\,\mathrm{d}z\qquad(2.38)$$

and

$$\phi(z, t) = h(z)\phi_0(z, t)\qquad(2.39)$$

given that $h(z)$ is a linear torsional shape function relating ϕ and ϕ_0. The reader is referred to the following references for a detailed treatment of the system kinematics required for the derivation of equations (2.37) and (2.38) (Cartmell, 1984; Bux and Roberts, 1986; Cartmell and Roberts, 1988). B_1 is a constant for the particular system and may be evaluated. Therefore the system kinetic energy is now,

$$T = \tfrac{1}{2}m_0(\dot{u}_0{}^2 + \dot{w}_0{}^2 + [\dot{v}_0 + \dot{W}]^2) - \tfrac{1}{2}I_0\dot{\phi}_0{}^2.\qquad(2.40)$$

This should be evident from consideration of Fig. (2.5(d)).

The strain energy is extended from equation (2.24) to include that due to torsion, giving

$$U = \tfrac{1}{2}EI_y\left[\int_0^l (f'')^2\,\mathrm{d}z\right]u_0{}^2 + \tfrac{1}{2}cGJ\left[\int_o^l (h')^2\,\mathrm{d}z\right]\phi_0{}^2,\qquad(2.41)$$

where c is a constant which accounts for the non-circular cross-section of the beam. Once again we will ignore the gravitational term. The contraction displacement w_0 is as in equation (2.26). Once again the method of Lagrange is applied, the energy expressions are differentiated appropriately (in the context of coordinates u_0 and ϕ_0) and the governing equations eventually emerge. Nonlinear (cubic) terms are evident once again, due to w_0 and v_0, and here too we discount them in an attempt to preserve linearity. The resulting equations of motion are

$$\ddot{u}_0 + 2\xi_b\omega_b\dot{u}_0 + \omega_b{}^2 u_0 + B_1\ddot{W}\phi_0 = 0, \qquad (2.42)$$

$$\ddot{\phi}_0 + 2\xi_t\omega_t\dot{\phi}_0 + \omega_t{}^2\phi_0 + \frac{m_0}{I_0} B_1\ddot{W}u_0 = 0. \qquad (2.43)$$

Coordinate coupling is strong in both equations (parametric excitation terms) showing that for this particular configuration motion in one mode requires motion in the order for instability to develop.

2.2.3 A note on the ordering of terms

The full complement of terms (not given here) for equations (2.36), (2.42) and (2.43) includes many that will not be strongly contributory to the resonance of interest. In fact some will have no appreciable effect whatsoever, and therefore we require a method of ordering which enables the relative importance of terms to be stated mathematically. Unfortunately there is no clear ruling that can be stated here, and the researcher is left to perform a somewhat *ad hoc* selection of terms that seem obviously significant (or 'strong') and those that appear to be less so. Equation (2.35) can be used to illustrate the problem, and so if we examine this we see that the first term $m_0\ddot{u}_0$ (the inertia term) is linear. It is fundamental to the motion of the system and is considered therefore to be 'strong'. As a result we do not need to interfere in any way with it. This is similarly true for the fourth term (which, when divided by m_0, comes out as $\omega^2 u_0$ – the stiffness term). These two terms then could be regarded as being of zeroth order (unaltered) and we could, for the sake of mathematical completeness, rewrite them as $\varepsilon^0\ddot{u}_0$ (m_0 dividing out through the terms), and $\varepsilon^0\omega^2 u_0$ respectively, where ε is a small, arbitrary, parameter. These terms form the linear homogeneous part of the problem. On the next level we consider those terms that dictate the flow of excitational energy through the system; these are the parametric excitation term and a damping term (which is not yet present in equation (2.35)). It is usually sufficiently representative to insert a linear viscous type of damping term into the derived equation of the form $c\dot{u}_0$, which becomes $2\xi\omega\dot{u}_0$ on division through the equation by m_0. Therefore we write these terms to first-order ε. In the case of the damping term a new damping ratio ζ is introduced, where,

$$\xi = \varepsilon\zeta. \qquad (2.44)$$

The numerical significance of ζ is unclear because ε is not specified, however it will be possible to return to ξ (the real physical parameter) later in the analysis, thus avoiding the problem of a lack of numerical definition for ζ.

The parametric excitation term is $B\ddot{W}u_0$ and it is convenient to work on this a little more so that it emerges, after ordering, in a useful form.

The excitation applied to the base of the beam is in the form of an acceleration, $\ddot{W}$, where

$$W(t) = W_0 \cos \Omega t \tag{2.45}$$

for example. Therefore we have

$$\ddot{W} = -W_0 \Omega^2 \cos \Omega t \tag{2.46}$$

enabling the parametric term to be written more fully

$$B\ddot{W}u_0 = (-BW_0\Omega^2 \cos \Omega t)u_0. \tag{2.47}$$

Since we have decided to consider this a first-order ε term it can be written in the following form

$$B|W_0\Omega^2| \cos \Omega t . u_0 = \varepsilon f \cos \Omega t u_0, \tag{2.48}$$

where f is a new excitation parameter with no direct defined numerical significance. Like ζ it will be restated as the physical parameter later in the analysis. For the meantime we have

$$BW_0\Omega^2 = \varepsilon f. \tag{2.49}$$

The remaining terms in equation (2.35) (second and third), are non-linear and essentially cubic in form. Their contribution to the principal parametric resonance condition is likely to be negligible (cubics in whatever form are not obviously resonant to principal parametric resonance, although a sensitive higher-order analysis could reveal some small contributory modifications arising as a result of their inclusion), and so they are regarded as small terms, in a relative sense. This 'smallness' is accentuated by the presence of derivatives within the cubics. For these reasons terms of this sort can be relegated to second-order ε; we could accomplish this by introducing a redefined system constant b such that

$$B = (\varepsilon b). \tag{2.50}$$

Now equation (2.35) may be restated in its ordered form

$$\varepsilon^0 \ddot{u}_0 + \varepsilon^2 b^2 u_0^2 \ddot{u}_0 + \varepsilon^2 b^2 u_0 \dot{u}_0^2 + \varepsilon^1 2\zeta\omega\dot{u}_0 + \varepsilon^0 \omega^2 u_0$$
$$- \varepsilon^1 f \cos \Omega t . u_0 = 0. \tag{2.51}$$

Note that the equation contains the added damping term and that m_0 has been divided out. This equation is now in a form suitable for further analysis, perhaps by a pertubation method or an averaging technique. The important point to recognize is that we have not formally altered the underlying structure of the derived equation (2.35) (other than by adding in the damping term), and that we could choose to proceed either with a simplified linearized treatment or a more complex, and more informative, nonlinear analysis. It should also be recognized that the preceding ordering scheme is by no means definitive. There is no rationale for ordering of terms in equations such as these other than perhaps on the basis outlined.

However, it appears, in the light of some experience, that the scheme described is generally suitable for this type of problem. Different problems with different physical origins might well require some other ordering priority. Caution should be exercised in the modification of governing equations and the reader is advised to investigate other examples besides those given within this book, in particular Nayfeh and Mook (1979).

Returning to equation (2.51) the 'strength' of each term depends closely on the order of ε. Since ε is defined as being small ($\varepsilon \ll 1$) then the lower order terms will be 'stronger' than those of higher-order ε.

Governing equations are frequently rearranged so as to non-dimensionalize each term. This has definite advantages when equations of the form of (2.42) and (2.43) (combination resonance problem) are involved in that the two parametric excitation terms, $B_1 \ddot{W} \phi_0$ and $(m_0/I_0) B_1 \ddot{W} u_0$, for example, can be tidied up. This non-dimensionalization is particularly useful when performed in conjunction with the introduction of the ordering parameter ε. Thus the algebra which follows in the later analysis will be helpfully reduced.

Conversely there seems little point in non-dimensionalizing equation (2.51) since it is a naturally well structured equation – especially so in its linearized form.

Taking equations (2.42) and (2.43) we introduce an arbitrary reference displacement b (this has length units) and define a new bending coordinate so that

$$X = \frac{u_0}{b}. \tag{2.52}$$

Similarly for the torsional coordinate we define the following

$$Y = \frac{\phi_0 a}{b}, \tag{2.53}$$

where

$$a = \sqrt{\frac{I_0}{m_0}} \left(\text{units:} \left(\frac{ML^2}{M} \right)^{1/2} = L \right). \tag{2.54}$$

Therefore X and Y are non-dimensionalized coordinates (noting that ϕ_0 in radians is itself dimensionless) but we need to be consistent in our transformation of both coordinates, hence the units of Y are

$$\frac{1 \cdot (ML^2/M)^{1/2}}{L} = 1 \qquad \text{(i.e. a non-dimensionalized coordinate)}.$$

So we now have

$$\ddot{X} + 2\xi_b\omega_b\dot{X} + \omega_b{}^2X - W_0\Omega^2B_1\cos\Omega t . Y\left(\sqrt{\frac{m_0}{I_0}}\right) = 0, \qquad (2.55)$$

$$\ddot{Y} + 2\xi_t\omega_t\dot{Y} + \omega_t{}^2Y - W_0\Omega^2B_1\left(\frac{m_0}{I_0}\right)\cos\Omega t . X\left(\sqrt{\frac{I_0}{m_0}}\right) = 0. \qquad (2.56)$$

Examination of the parametric excitation terms in both equations will reveal that they are now identical (due to the definition of Y).

The ordering scheme discussed above can now be adopted and to this end we can, for simplicity, define ε as

$$\varepsilon = W_0\Omega^2R_1\sqrt{\frac{m_0}{I_0}}. \qquad (2.57)$$

The damping terms are ordered to first-order ε such that

$$\xi_b = \varepsilon\zeta_b \qquad \text{and} \qquad \xi_t = \varepsilon\zeta_t. \qquad (2.58)$$

The use of ε (which is now effectively the excitation parameter) to order the damping does not cause difficulties because $\zeta_{b,t}$ have no clear numerical significance, and are merely convenient descriptors for the damping in the analysis which follows. So finally the two equations of motion are

$$\ddot{X} + 2\varepsilon\zeta_b\omega_b\dot{X} + \omega_b{}^2X - \varepsilon\cos\Omega t . Y = 0, \qquad (2.59)$$

$$\ddot{Y} + 2\varepsilon\zeta_t\omega_t\dot{Y} + \omega_t{}^2Y - \varepsilon\cos\Omega t . X = 0. \qquad (2.60)$$

Another transformation technique that can be useful is the shifting of the time-scale t to some non-dimensionalized scale τ. The need for this approach depends very much on the nature of the equations and will be examined further in a later Section 3.2.2(b).

2.3 ANALYSIS OF GOVERNING EQUATIONS

In this section equations (2.59) and (2.60) are treated using two different methods of analysis, firstly by the method of multiple scales and then by the method of harmonic balance. It will be shown that both methods lead to the same result. The Introduction and Section 2.1 of this chapter emphasized the fact that solutions to equations such as (2.59) and (2.60) are unbounded with time, and that the main task is to determine where (in the two parameter plane) non-zero solutions are likely to be found rather than to attempt to find the solutions themselves. Therefore the analyses which follow lead to an expression for the detuning parameter (which will be denoted by $\varepsilon\sigma$). This can be used to plot the transition curve.

2.3.1 Application of the method of multiple scales

This is a pertubation method in which the variables, or coordinates, are written in series form so that (on back substitution into the governing equations) a set of pertubation equations can be constructed by grouping terms of like order ($\varepsilon^0, \varepsilon^1, \varepsilon^2, \ldots$). Successive solution of each perturbation equation allows a final general solution to be built up. We apply the method in this form except that an explicit solution is not sought. The principal difference between this and other pertubation methods is that the time (independent variable) is represented by independent time-scales of the form

$$T_n = \varepsilon^n t, \qquad n = 0, 1, 2 \ldots .. \tag{2.61}$$

Variables X and Y are written in asymptotic series form (theoretically up to any order but which, in practice, need to be truncated to some manageable level). Taking the series up to and including first-order ε will be shown to be sufficient for this example, therefore we have

$$X = X_0 + \varepsilon X_1 + \ldots; \qquad Y = Y_0 + \varepsilon Y_1 + \ldots .. \tag{2.62}$$

The series are assumed to be uniformly convergent. The derivatives are also treated in this way, giving, up to first order

$$\frac{d}{dt} = \frac{\partial}{\partial t_0} + \varepsilon \frac{\partial}{\partial T_1} + \ldots;$$

$$\frac{d^2}{dt^2} = \left(\frac{\partial}{\partial T_0}\right)^2 + 2\varepsilon\left(\frac{\partial}{\partial T_0}\right)\left(\frac{\partial}{\partial T_1}\right) + \ldots \tag{2.63}$$

in which it is more convenient to adopt D operator notation so that

$$\frac{d}{dt} = D_0 + \varepsilon D_1 + \ldots; \qquad \frac{d^2}{dt^2} = D_0{}^2 + 2\varepsilon D_0 D_1 + \ldots . \tag{2.64}$$

The forms in equations (2.62), (2.64) are substituted into governing equations (2.59), (2.60) respectively and the equations are arranged in such a manner that terms to like order of ε are grouped together, therefore to ε^0 we have,

$$\varepsilon^0: \quad D_0{}^2 X_0 + \omega_b{}^2 X_0 = 0, \tag{2.65}$$

$$D_0{}^2 Y_0 + \omega_t{}^2 Y_0 = 0, \tag{2.66}$$

and to order ε^1

$$\varepsilon^1: \quad D_0{}^2 X_1 + \omega_b{}^2 X_1 = -2D_0 D_1 X_0 - 2\zeta_b \omega_b D_0 X_0 + Y_0 \cos \Omega t, \tag{2.67}$$

$$D_0{}^2 Y_1 + \omega_t{}^2 Y_1 = -2D_0 D_1 Y_0 - 2\xi_t \omega_t D_0 Y_0 + X_0 \cos \Omega t. \tag{2.68}$$

Equations (2.65) and (2.66) are sometimes referred to as the zeroth-order perturbation equations and their solutions are immediately apparent, noting that the method favours the following exponential forms

(where the bars denote complex conjugates)

$$X_0 = A(T_1) \exp(i\omega_b T_0) + \bar{A}(T_1) \exp(-i\omega_b T_0), \qquad (2.69)$$

$$Y_0 = B(T_1) \exp(i\omega_t T_0) + \bar{B}(T_1) \exp(-i\omega_t T_0). \qquad (2.70)$$

The zeroth-order solutions may now be substituted into the first-order perturbation equations leading to the following for equation (2.67)

$$D_0^2 X_1 + \omega_b^2 X_1 = \exp(i\omega_b T_0)[(B/2) \exp i(\omega_t + \Omega - \omega_b)T_0$$

$$+ (\bar{B}/2) \exp i(\Omega - \omega_t - \omega_b)T_0 - i2\omega_b D_1 A$$

$$- i2\zeta_b \omega_b^2 A] + \text{C.C.} \qquad (2.71)$$

since we can express $\cos \Omega t$ as

$$\cos \Omega t = \tfrac{1}{2}(\exp(i\Omega T_0) + \exp(-i\Omega T_0)); \qquad t \simeq T_0. \qquad (2.72)$$

Note that C.C. represents the complex conjugates of the preceding terms.

The next stop is to consider how equation (2.71) should be treated so that its particular solution does not invalidate the important underlying necessity for uniformity in the expansion. This uniformity is not likely to occur if certain terms are present in the first-order pertubation equation – these terms will in turn give rise to secular terms in the particular solution to this equation if measures of some sort are not taken. The danger then is that the secular terms will create a disproportionate increase in the relative magnitude of the additional correction generated by solving this order of perturbation equation. Such correction terms would of course normally be combined with the solution to lower-order pertubation equations to form the general solution to the problem, thus the first-order correction should be of relatively small magnitude compared with the zeroth-order solution. Therefore terms appearing in the higher-order perturbation equations (in this case we are only going as far as first order), which can be identified as ones that will give rise to secular terms in the particular solution to that equation must be eliminated.

The easiest way of tackling this problem and in so doing precluding the generation of secular terms (without actually going to the trouble of trying to work out the particular solution) is to 'take out' the resonant exponent term – in this case $\exp(i\omega_b T_0)$ as shown in equation (2.71) and then to set the contents of the bracket to zero. At this stage any non-resonant, or fast, terms within the bracket are simply ignored. These would be easily recognizable because they would contain 'fast' or larger exponents, i.e. $\exp(i2\omega_b T_0)$, $\exp(i3\omega_b T_0)$ etc. Therefore equating the bracketed terms to zero gives

$$\frac{B}{2} \exp i(\omega_t + \Omega - \omega_b) T_0 + \frac{\bar{B}}{2} \exp i(\Omega - \omega_t - \omega_b) T_0$$

$$- i2\omega_b D_1 A - i2\zeta_b \omega_b^2 A = 0. \tag{2.73}$$

The two final terms of equation (2.73) are resonant and capable of generating secular terms in the subsequent particular solution. It is therefore absolutely appropriate that they appear in equation (2.73), however as they stand the first two terms might, or might not, be resonant. We need to identify Ω in terms of ω_b and ω_t in some way.

Equation (2.20) describes a general combination resonance involving two natural frequencies, and in this example we will try to investigate the sum type combination for which $m = n = 1$. Equation (2.20) now becomes

$$\Omega = (\omega_b + \omega_t) + \varepsilon\rho. \tag{2.74}$$

Substitution for Ω in the exponent of the first term of equation (2.73) yields, $(B/2) \exp i(2\omega_t + \varepsilon\rho)$; this is demonstrably a fast non-resonant term and will therefore be neglected. The second term of equation (2.73) reduces to $(\bar{B}/2) \exp i(\varepsilon\rho) T_0$, and since $\varepsilon\rho$ is defined as small this term is resonant enough to be considered in the same light as the $i2\omega_b D_1 A$ and $i2\zeta_b \omega_b^2 A$ terms – ones that will generate secular terms and thus allow X_1 to grow disproportionately large. The result of applying equation (2.74) is now

$$\frac{\bar{B}}{2} \exp i(\varepsilon\rho) T_0 - i2\omega_b D_1 A - i2\zeta_b \omega_b^2 A = 0. \tag{2.75}$$

This equation has grown out of the need to protect the uniformity of the expansions of X and Y and it is an interesting and powerful feature of the method that it provides sufficient information in itself to complete the analysis. So we do not, in this example, require to go back to the first-order perturbation equation (2.71) and formally solve it.

If we were interested in obtaining A or B (which we are not in this case, merely requiring an analytic detuning parameter) an equation (or equations) such as equation (2.75) could be integrated numerically (or treated approximately by analysis if straightforward enough), to give the envelope of the oscillatory solution(s) to the governing equations. Returning to equation (2.75) it is helpful to state amplitudes A and B in their polar forms, introducing phase angles α and β respectively, therefore we return to the zeroth-order solutions, equations (2.69) and (2.70) (where complex amplitudes A and B were originally introduced), and write these in an alternative form

$$X_0 = a \cos(\omega_b T_0 + \alpha), \tag{2.76}$$

$$Y_0 = b \cos(\omega_t T_0 + \beta), \tag{2.77}$$

where α, β are phase angles.

As

$$\cos p = \tfrac{1}{2}[\exp(\mathrm{i}p) + \exp(-\mathrm{i}p)] \qquad (2.78)$$

then

$$X_0 = a\cos(\omega_b T_0 + \alpha) = a\{\tfrac{1}{2}[\exp\mathrm{i}(\omega_b T_0 + \alpha) + \exp\mathrm{i}(-\omega_b T_0 - \alpha)]\}. \qquad (2.79)$$

Therefore,

$$X_0 = a\{\tfrac{1}{2}[\exp\mathrm{i}(\omega_b T_0)\exp.\mathrm{i}\alpha + \exp\mathrm{i}(-\omega_b T_0).\exp\mathrm{i}(-\alpha)]\}. \qquad (2.80)$$

Thus by examining equations (2.69) and (2.80) it can be seen that the complex amplitude A is defined by

$$A = \frac{a}{2}\exp\mathrm{i}\alpha; \qquad \bar{A} = \frac{a}{2}\exp\mathrm{i}(-\alpha). \qquad (2.81)$$

Similarly for complex amplitude B

$$B = \frac{b}{2}\exp\mathrm{i}\beta; \qquad \bar{B} = \frac{b}{2}\exp\mathrm{i}(-\beta). \qquad (2.82)$$

These are the polar forms of A and B.

The second term of equation (2.75) contains the following partial derivative

$$D_1 A = \frac{\partial A}{\partial T_1}; \qquad \text{where } A = A(T_1). \qquad (2.83)$$

Substitution of the polar form of A into equation (2.83) gives

$$D_1 A = \frac{a'}{2}\exp\mathrm{i}\alpha + \frac{a}{2}\mathrm{i}\alpha'\exp\mathrm{i}\alpha; \qquad (\alpha = \alpha(T_1); a = a(T_1)), \qquad (2.84)$$

where the prime denotes differentiation with respect to time-scale T_1. Making the necessary substitutions for A and B in equation (2.75) produces

$$\frac{b}{4}\exp\mathrm{i}(\rho T_1 - \alpha - \beta) - \mathrm{i}\omega_b a' + \omega_b a\alpha' - \mathrm{i}\zeta_b\omega_b{}^2 a = 0. \qquad (2.85)$$

This equation contains both real and imaginary terms so it can be split into two more equations on this basis

$$\text{Re:} \quad \frac{b}{4}\cos(\rho T_1 - \alpha - \beta) + \omega_b a\alpha' = 0, \qquad (2.86)$$

$$\text{Im:} \quad \frac{b}{4}\sin(\rho T_1 - \alpha - \beta) - \omega_b a' - \zeta_b\omega_b{}^2 a = 0. \qquad (2.87)$$

Noting that $\rho T_1 = \varepsilon\rho T_0$ from equation (2.61).

An identical procedure, but operating on equation (2.68) will yield two other equations identical in form

$$\text{Re:} \quad \frac{a}{4} \cos (\rho T_1 - \alpha - \beta) + \omega_\text{t} b \beta' = 0, \tag{2.88}$$

$$\text{Im:} \quad \frac{a}{4} \sin (\rho T_1 - \alpha - \beta) = \omega_\text{t} b' - \zeta_\text{t} \omega_\text{t}^2 b = 0. \tag{2.89}$$

Equations (2.86) to (2.89) are first-order differential equations with the amplitudes a, and b, and the phase α, and β, as dependent variables, and slow time T_1 as the independent variable, thus

$$a = a(T_1); \qquad b = b(T_1); \qquad \alpha = \alpha(T_1); \qquad \beta = \beta(T_1).$$

There are now two possible approaches either of which could be taken; one is numerical integration which could be used for investigating a and b, the other being a further analysis by which we attempt to derive an analytical expression for $\varepsilon \rho$. This latter course inevitably will involve some approximative work on equations (2.88) and (2.89) in order to achieve the desired results. We will attempt to show that a first-order multiple scales analysis can provide a useful approximation by which the transition curve may be plotted, hence the second of the two options given above.

Since equations (2.86) and (2.89) contain slowly varying parameters (the equations are with respect to time-scale T_1 rather that T_0 or t) the assumption that the amplitudes are virtually static in slow time is a reasonable one. Thus

$$a' = b' \simeq 0. \tag{2.90}$$

We also eliminate the explicit presence of time (in whatever time-scale form it is represented) and in so doing obtain an autonomous system by stating

$$\rho T_1 - \alpha - \beta = \Psi. \tag{2.91}$$

It should be noted that equation (2.90) is not strictly true, particularly over many real time units (t or T_0, since $t \simeq T_0$) and that a and b would be seen to increase in an unbounded manner over a long real-time period. However it will suffice for the purposes of this analysis.

Applying equation (2.90) and (2.91) to equations (2.88) to (2.89) we obtain

$$(b/4) \cos \Psi + \omega_\text{b} a \alpha' = 0, \tag{2.92}$$

$$(b/4) \sin \Psi - \zeta_\text{b} \omega_\text{b}^2 a = 0, \tag{2.93}$$

$$(a/4) \cos \Psi + \omega_\text{t} b \beta' = 0, \tag{2.94}$$

$$(a/4) \sin \Psi - \zeta_\text{t} \omega_\text{t}^2 b = 0. \tag{2.95}$$

The slowly varying phase angles α' and β' are still present, however, rather than ruling these terms out (and losing equations (2.92) and (2.94) in the process) it is helpful to consider analytical substitutions in the following manner. We achieve this by reconsidering equation (2.91) and arguing that the autonomous system angle Ψ can be regarded in the same way as the amplitudes, i.e. it is very slowly changing, thus,

$$\Psi' \simeq 0; \qquad \rho - \alpha' - \beta' \simeq 0. \tag{2.96}$$

Isolation of α' in equation (2.96) and substitution into equation (2.92) provides an expression for β'

$$\beta' = \rho\bigg/\bigg[1 + \frac{\omega_t \zeta_b}{\omega_t \zeta_t}\bigg]. \tag{2.97}$$

Similarly from equation (2.94) we obtain α'

$$\alpha' = \rho\bigg/\bigg[1 + \frac{\omega_t \zeta_t}{\omega_b \zeta_b}\bigg]. \tag{2.98}$$

Equations (2.92) to (2.95) are sometimes called the solvability equations, and α' and β' having been identified, they are now at a point where further manipulation will be purely algebraic. We can now proceed to solve for ρ (hence $\varepsilon\rho$).

There are two cases to consider here:

1. case when $\zeta_b = \zeta_t = 0$ (no damping);
2. case when $\zeta_b \neq 0$; $\zeta_t \neq 0$ (damped system).

In order to investigate these cases we substitute for α' and β' and manipulate equations (2.92) to (2.95) so that amplitudes a and b are eventually no longer explicit. This leads to

$$\frac{b^2}{16a^2} = \frac{\omega_b^2 \rho^2}{[1 + \omega_t \zeta_t/\omega_b \zeta_b]^2} + \omega_b^4 \zeta_b^2, \tag{2.99}$$

$$\frac{a^2}{16b^2} = \frac{\omega_t^2 \rho^2}{[1 + \omega_b \zeta_b/\omega_t \zeta_t]^2} + \omega_t^4 \zeta_t^2. \tag{2.100}$$

It is now clear that a and b will substitute out, and further algebraic manipulation will lead to a biquadratic in ρ

$$\rho^4 + 2\rho^2 [\omega_b \zeta_b + \omega_t \zeta_t]^2$$
$$+ \bigg[\omega_b^2 \omega_t^2 \zeta_b^2 \zeta_t^2 - \frac{1}{16^2 \omega_b^2 \omega_t^2}\bigg]\bigg[1 + \frac{\omega_b \zeta_b}{\omega_t \zeta_t}\bigg]^2\bigg[1 + \frac{1\omega_t \zeta_t}{\omega_b \zeta_b}\bigg]^2 = 0.$$

$$\tag{2.101}$$

The zero damping case reduces equation (2.101) to

$$\rho^4 - \frac{1}{(16\omega_b \omega_t)^2} \tag{2.102}$$

and so,

$$\rho = \pm \frac{1}{4\sqrt{\omega_b \omega_t}}. \tag{2.103}$$

Hence substitution of this into the combination resonance of equation (2.74) provides

$$\Omega = \omega_b + \omega_t \pm \frac{\varepsilon}{4\sqrt{\omega_b \omega_t}}. \tag{2.104}$$

For the second case (non-zero damping) it is necessary to solve for ρ in equation (2.101) with $\zeta_b \neq 0$, $\zeta_b \neq 0$. This results in the following expression

$$\rho = \frac{[1 + \omega_t \zeta_t / \omega_b \zeta_b]}{2\sqrt{\omega_t \zeta_t / \omega_b \zeta_b}} \sqrt{\frac{1}{4\omega_b \omega_t} - 4\omega_b \omega_t \zeta_b \zeta_t}. \tag{2.105}$$

And so, multiplying through by ε,

$$\varepsilon\rho = \frac{[1 + \omega_t \xi_t / \omega_b \xi_b]}{2\sqrt{\omega_t \xi_t / \omega_b \xi_b}} \sqrt{\frac{\varepsilon^2}{4\omega_b \omega_t} - 4\omega_b \omega_t \xi_b \xi_t}. \tag{2.106}$$

This now defines the detuning parameter in terms of the original physical parameters (notably damping ξ_b, ξ_t) and incorporates ε which was originally defined as an excitation function as well as a perturbation parameter. Equation (2.106) for the damping problem could be substituted into the combination resonance equation (2.74) and the transition

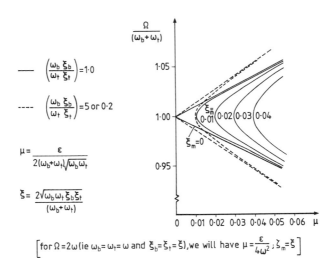

$$\left(\frac{\omega_b}{\omega_t} \frac{\xi_b}{\xi_t}\right) = 1.0$$

$$\left(\frac{\omega_b}{\omega_t} \frac{\xi_b}{\xi_t}\right) = 5 \text{ or } 0.2$$

$$\mu = \frac{\varepsilon}{2(\omega_b + \omega_t)\sqrt{\omega_b \omega_t}}$$

$$\xi = \frac{2\sqrt{\omega_b \omega_t} \xi_b \xi_t}{(\omega_b + \omega_t)}$$

$$\left[\text{for } \Omega = 2\omega \text{ (ie } \omega_b = \omega_t = \omega \text{ and } \xi_b = \xi_t = \xi\text{), we will have } \mu = \frac{\varepsilon}{4\omega^2}; \xi_m = \xi\right]$$

Figure 2.6 Stability chart showing the transient curve boundaries for the bending/torsion combination resonance. The effects of system damping are highlighted. Note that this is also a master curve for principal parametric resonance – see the annotation on the figure (after Dugundji and Mukhopadhyay, 1973).

curve calculated for various physical parameter values. This is depicted in Fig. 2.6. The effects of damping on the shape of the curve are to be highlighted in Section 2.3.3.

2.3.2 Application of the method of harmonic balance

This technique can also be used to treat the problem in hand, the reader is referred to two published research papers for full theoretical and experimental details (Yamamoto and Saito, 1970; Dugundji and Mukho-padhyay, 1973). For the sake of consistency we will persist with the established nomenclature of the previous section. The method requires the assumption of solution forms to equations (2.59) and (2.60) and so we can therefore state the following

$$X(t) = a_1 \sin \omega_1 t + b_1 \cos \omega_1 t, \qquad (2.107)$$

$$Y(t) = a_2 \sin \omega_2 t + b_2 \cos \omega_2 t, \qquad (2.108)$$

where a_i and b_i are constants and ω_1 and ω_2 are yet undetermined frequencies. We also state that

$$\Omega = \omega_1 + \omega_2 \qquad (2.109)$$

although it should be noted that we are not yet attempting to align $\omega_{1,2}$ with $\omega_{b,t}$ respectively, but that this will indeed come out towards the end of the analysis.

Equations (2.107), (2.108) and (2.109) are substituted into equations (2.59) and (2.60) with the result that all terms in each equation contain either $\cos \omega_{1,2} t$ or $\sin \omega_{1,2} t$. The method requires these terms to be grouped accordingly and after some algebra and simplification we arrive at

$$\sin \omega_1 t : (\omega_b^2 - \omega_1^2)a_1 - c_1\omega_1 b_1 + \varepsilon(a_2/2) = 0, \qquad (2.110)$$

$$\cos \omega_1 t : c_1\omega_1 a_1 + (\omega_b^2 - \omega_1^2)b_1 - \varepsilon (b_2/2) = 0, \qquad (2.111)$$

$$\sin \omega_2 t : \varepsilon(a_1/2) + (\omega_t^2 - \omega_2^2)a_2 - c_2\omega_2 b_2 = 0, \qquad (2.112)$$

$$\cos \omega_2 t : -\varepsilon(b_1/2) + c_2\omega_2 a_2 + (\omega_t^2 - \omega_2^2)b_2 = 0. \qquad (2.113)$$

Note that the ordered damping terms of equation (2.59) and (2.60) have been replaced with the following

$$2\varepsilon\zeta_b\omega_b = c_1; \qquad 2\varepsilon\zeta_t\omega_t = c_2. \qquad (2.114)$$

Equations (2.110) to (2.113) represent the harmonic balancing of equations (2.59) and (2.60) and subsequent work is mainly concerned with manipulation of these into a usable form. It is possible to reduce the harmonically balanced equations down further by assuming the following forms

$$a_2 = \gamma a_1; \qquad b_2 = \gamma b_1, \tag{2.115}$$

$$(\omega_b{}^2 - \omega_1{}^2) = \gamma^2(\omega_t{}^2 - \omega_2{}^2), \tag{2.116}$$

$$\omega_1 c_1 = \gamma^2 \omega_2 c_2. \tag{2.117}$$

Therefore equations (2.110) to (2.113) simplify to give

$$(\omega_b{}^2 - \omega_1{}^2 + (\gamma/2)\,\varepsilon)a_1 = \omega_1 c_1 b_1 = 0, \tag{2.118}$$

$$\omega_2 c_2 a_1 + (\omega_t{}^2 - \omega_2{}^2 - (\varepsilon/2\gamma))b_1 = 0. \tag{2.119}$$

These equations could be rewritten in matrix form

$$\begin{bmatrix} (\omega_b{}^2 - \omega_1{}^2 + (\gamma/2)\varepsilon) & -\omega_1 c_1 \\ \omega_2 c_2 & (\omega_t{}^2 - \omega_2{}^2) - (\varepsilon/2\gamma) \end{bmatrix} \begin{bmatrix} a_1 \\ b_1 \end{bmatrix} = 0. \tag{2.120}$$

And so the transition curve could be obtained by setting the determinant of equation (2.120) to zero, but before this is done the unspecified frequencies ω_1 and ω_2 must be found, and γ eliminated by substitution. The frequencies pose little problem, and we reason that the coupling between equations (2.59) and (2.60) allows us to equate the system frequencies with those of the assumed solutions

$$\omega_1 \simeq \omega_b; \qquad \omega_2 \simeq \omega_t. \tag{2.121}$$

Therefore,

$$(\omega_b - \omega_1)^2 \simeq 0; \qquad (\omega_t - \omega_2)^2 \simeq 0, \tag{2.122}$$

and so

$$\omega_b{}^2 - \omega_1{}^2 \simeq 2\omega_b(\omega_b - \omega_1), \tag{2.123}$$

$$\omega_t{}^2 - \omega_2{}^2 \simeq 2\omega_t(\omega_t - \omega_2). \tag{2.124}$$

Substitution of equations (2.116) into equations (2.123), (2.124) and bringing in equation (2.109) leads to the following

$$\omega_1 = \frac{1}{1 + c_2/c_1} \left[\Omega + \frac{c_2}{c_1} \omega_b - \omega_t \right], \tag{2.125}$$

$$\omega_2 = \frac{c_2}{c_1(1 + c_2/c_1)} \left[\Omega - \omega_b + \frac{c_1}{c_2} \omega_t \right]. \tag{2.126}$$

If we now set the determinant of equation (2.120) to zero and substitute for $\omega_{1,2}$ as required, from equations (2.125) and (2.126) the stability boundary expression begins to emerge, therefore

$$\frac{\Omega}{(\omega_b + \omega_t)} = 1 \pm \frac{[1 + c_2/c_1]}{2\sqrt{c_2/c_1}} \sqrt{\frac{\varepsilon^2}{4(\omega_b + \omega_t)^2 \omega_b \omega_t} - \frac{c_1 c_2}{(\omega_b + \omega_t)^2}}. \tag{2.127}$$

Finally by substituting for c_1 and c_2 in the manner of equation (2.114), where $\xi_{b,t} = \varepsilon\zeta_{b,t}$, we have

$$\Omega = \omega_b + \omega_t \pm \frac{[1 + \omega_t\xi_t/\omega_b\xi_b]}{2\sqrt{\omega_t\xi_t/\omega_b\xi_b}} \sqrt{\frac{\varepsilon^2}{4\omega_b\omega_t} - 4\omega_b\omega_t\xi_b\xi_t}. \quad (2.128)$$

If we now return to the previous section and substitute for $\varepsilon\rho$ (of equation (2.106)) into equation (2.74) it is apparent that the methods of multiple scales and harmonic balance can give identical results for this particular example.

2.3.3 The stability boundary

Equation (2.128) is shown plotted in Fig. 2.6 in such a manner that the frequency ratio (another way of representing the detuning parameter) is the ordinate and an excitation parameter μ is on the abscissa. We define μ in the following way

$$\mu = \frac{\varepsilon}{2\sqrt{\omega_b\omega_t}\ (\omega_b + \omega_t)}. \quad (2.129)$$

This particular stability boundary was previously presented in this form (Dugundji and Mukhopadhyay, 1973) and was, in turn, derived from earlier original work (Yamamoto and Saito, 1970).

The most important aspect of Fig. 2.6 is the tendency for the unstable region to pull back from the frequency axis with increasing damping ξ_m, where this is a composite term containing both bending and torsional damping terms, defined in Fig. 2.6. The effect of changing the ratio of ξ_b to ξ_t is shown by the dotted lines, therefore if either damping value is reduced to one fifth (say) of the other the overall effect is for the curve to widen out. We can preserve this ratio but increase ξ_m in an overall sense and as a result observe the pulling away of the wider, dotted, curve from the frequency axis. Increasing the damping, ξ_m, eventually brings about the case for which

$$\xi_m > \mu \quad (2.130)$$

and at this point the unstable region disappears altogether.

Therefore as long as inequality (2.130) does not hold, the support motion of the simple structure of Fig. 2.5(d) will drive the system into a sum-type combination instability for excitational frequency – acceleration levels which fall within the predicted zone given in Fig. 2.6. The magnitude of the response amplitudes, both in bending and torsion, will, at least theoretically, grow with time and become infinitely large. Of course this will not happen in practice because other limiting effects begin to take over as a result of the extremely large deflections. We examine non-linearities such as these in Section 3.1.2.

It is interesting to note that if we replace subscript b with subscript t (or vice versa) in equation (2.128) it reduces down to

$$\Omega = 2\omega_{b,t} \pm 2 \sqrt{\frac{\varepsilon^2}{16\omega_{b,t}^2} - \omega_{b,t}^2 \xi^2} \,. \tag{2.131}$$

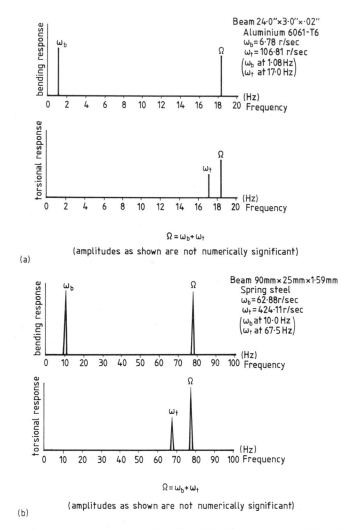

Figure 2.7 Frequency spectra for bending/torsion combination resonance of: (a) an aluminium cantilever; (b) a steel cantilever (after Dugundji and Mukhopadhyay, 1973).

This is the expression for the transition curve pertaining to the principal parametric resonance case. It can involve motion in either of the fundamental bending or torsion modes. So Fig. 2.6 is really a master plot for $\Omega \simeq 2\omega_b$; $\Omega \simeq 2\omega_t$; $\Omega \simeq \omega_b + \omega_t$.

A further modification is the case of the difference type combination instability which will result when the governing equations (2.59), (2.60) are antisymmetrically coupled, so generally we have,

$$\ddot{X} + 2\varepsilon\zeta_b\omega_b\dot{X} + \omega_b^2X \pm \varepsilon\cos\Omega\, t . Y = 0, \qquad (2.132)$$

$$\ddot{Y} + 2\varepsilon\zeta_t\omega_t\dot{Y} + \omega_t^2Y \mp \varepsilon\cos\Omega\, t . X = 0. \qquad (2.133)$$

Here the primary region occurs at

$$\Omega = (\omega_t - \omega_b) \pm \varepsilon\rho \qquad (2.134)$$

noting that this physical example will preclude $\Omega \simeq \omega_b - \omega_t$ since it is highly unlikely that $\omega_b > \omega_t$ will be the case. Equation (2.128) still holds except that $\omega_b + \omega_t$ is replaced by $\omega_t - \omega_b$. Therefore Figs. 2.6 is also a master plot for $\Omega \simeq \omega_t - \omega_b$.

Spectral plots and amplitude responses with time are given in Figs. 2.7

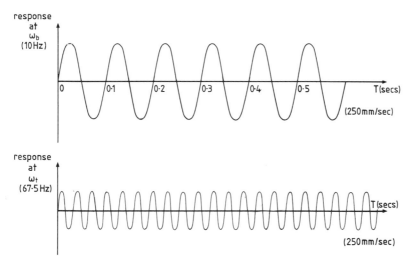

Beam 90mm×25mm×1·59mm steel, ω_b=62·83 r/sec (10·0 Hz), ω_t=424·11 r/sec (67·5 Hz)

Figure 2.8 Amplitude responses with time for bending/torsion combination resonance of a steel cantilever.

and 2.8. These are extracted from experimental work by the author and others. The non-synchronous nature of the responses with the excitations is clearly shown throughout.

2.4 FURTHER INVESTIGATIONS INTO PARAMETRIC PHENOMENA

There is now an immense body of literature dealing with a large variety of studies related to parametric phenomena. It seems prudent to categorize the areas of work in some way and the following descriptive headings are to be used here; fundamental studies, machine problems, structural elements and configurations, problems in applied physics (thermodynamics and electrical engineering for example).

The first category, fundamental studies, covers a vitally important area of research activity, where in the main, physical models are, initially at least, of secondary importance to the mathematical problems. There has been a large international effort in this area during the last few decades and, apart from its own inherent value, fundamental research ultimately leads the way towards the eventual solution of practical problems. We will discuss some important topics that have recently been highlighted in the literature, in particular, multifrequency parametric excitation, single fluctuating frequency parametric excitation, simultaneous forced and parametric problems, and higher-order combination resonances; refer to Section 2.4.1.

Investigations which might be appropriately categorized as machine problems comprise studies in the parametric vibration of pulley and belt systems, together with bandsaw instabilities, chain drives and belt tensioners. Certain examples will be discussed in Section 2.4.2.

Probably the largest category of applied parametric vibration problems is that of structures since this term covers beams, rods, plates, shells and various interacting configurations of these, and also structure/fluid interactions such as occur in rockets, missiles and satellites, as well as cases of general 'aerodynamic' interactions. Two structural problems are considered in Section 2.4.3.

The fourth category really covers any parametric problems that are modelled by physical effects not catered for in the previous two categories, notably in thermodynamics, an area of particular importance to mechanical engineers, and therefore suitable for inclusion in this book.

An example in this category can be found in Section 2.4.4.

2.4.1 Fundamental problems in parametric vibrations

Principal and first-order combination resonances have been cursorily examined in Sections 2.2 and 2.3 using the base excited cantilever structure as a basis for modelling. Whilst it is not entirely necessary to relate interesting governing equations of motion to physical systems, doing this almost always aids general understanding, particularly for the engineer. The counter to this is that conceptually difficult mathematical problems benefit from a full unambiguous treatment with minimal concessions to real world systems at first, so that a theoretical platform is developed after which modifications can be introduced and practical problems dealt with. Ignoring the needs of the real world allows unconstrained, theoretical research which might well generate answers to questions yet to be asked.

In addition to these two approaches we have the type of problem in which perhaps the structure is simplified but the form of excitation or damping accurately conforms to a practical situation, so that the close modelling of one problem aspect masks the necessary approximation in another – necessary because most vibration problems of any interest are inherently complicated, often making exact analysis and exact solution a near impossible achievement, with some form of approximation being vital in reaching an answer. In this section, problems which fall into one or other of the above types are discussed; it should be pointed out, however that the nonlinear topics are as far as possible avoided in the examples that follow.

(a) Multifrequency parametric excitation

In this problem the parametric excitation is regarded as a multifrequency function providing sinusoidal inputs with a constant frequency spacing. The system upon which this excitation acts is a single degree of freedom oscillator with damping, and we examine its behaviour when the difference between any two frequencies of the parametric excitation is in the vicinity of twice the natural frequency of the system (Othman, Watt and Barr, 1987). This is a case of principal parametric reasonance but under unusual conditions in which the difference between any two of the multiple input sinusoids is important. Thus the input frequencies might well be much larger than the system natural frequency yet still be capable of evoking a response under the stated resonance conditions (equation (2.137)). The system equation of motion is

$$\ddot{x} + 2\varepsilon\mu\omega\dot{x} + \omega^2 x + \varepsilon f(t)x = 0, \tag{2.135}$$

where ε, μ, ω are constants and the dots denote real time (t) differentiations. The parametric excitation $f(t)$ is given as the following generalized form

$$f(t) = \sum_{r=-n}^{r=n} \cos[(1 + r\alpha)t + \phi_r]; \qquad n = 1, 2, 3, \ldots \tag{2.136}$$

The input sinusoids have initial phase angles ϕ_r and the frequency spacing between them is constant, and denoted by α. The original authors of this work mention, as an application of the theory, that an excitation of the form of equation (2.136) could have been the cause of reported fractures in locomotive coupling rod problems (Den Hartog, 1934).

Since we are interested in the effects of differences between two frequencies, particuarly when close to twice the natural frequency, we have as a general resonance condition,

$$|r_1 - r_2| \, \alpha \simeq 2\omega, \tag{2.137}$$

where

$$|r_1 - r_2| = 1, 2, \ldots, 2n.$$

The methods of multiple scales is used up to and including second-order terms in the expansion, therefore,

$$x(t, \, \varepsilon) = x_0(T_0, \, T_1, \, T_2) + \varepsilon x_1(T_0, \, T_1, T_2) + \varepsilon^2 x_2(T_0, \, T_1, \, T_2) + \ldots. \tag{2.138}$$

To second order the derivatives are

$$\frac{\mathrm{d}}{\mathrm{d}t} = D_0 + \varepsilon D_1 + \varepsilon^2 D_2 + \ldots, \tag{2.139}$$

$$\frac{\mathrm{d}^2}{\mathrm{d}t^2} = D_0{}^2 + 2D_0 D_1 + \varepsilon^2(D_1{}^2 + 2D_0 D_2) + \ldots. \tag{2.140}$$

Substituting equations (2.136), (2.138), (2.139) and (2.140) into equation (2.135) leads to the following, ordered perturbation equations

$$\varepsilon^0 : D_0{}^2 x_0 + \omega^2 x_0 = 0, \tag{2.141}$$

$$\varepsilon^1 : D_0{}^2 x_1 + \omega^2 x_1 = -2D_0 D_1 x_0 - 2\mu\omega D_0 x_0$$

$$- x_0 \sum \cos\left[(1 + r\alpha)t + \phi_r\right], \tag{2.142}$$

$$\varepsilon^2 : D_0{}^2 x_2 + \omega^2 x_2 = -2D_0 D_1 x_1 - (D_1{}^2 + 2D_0 D_2)x_0 - 2\mu\omega D_1 x_0$$

$$-2\mu\omega D_0 x_1 - x_1 \sum_{r=-n}^{r=n} \cos\left[(1 + r\alpha)t + \phi_r\right]. \tag{2.143}$$

The solution to equation (2.141) is identical in form to that of equations (2.65, 2.66) therefore,

$$X_0 = A(T_1)\exp{(i\omega T_0)} + \bar{A}(T_1)\exp{(-i\omega T_0)}. \tag{2.144}$$

Substitution of solution (2.144) into equation (2.142) gives

$$D_0{}^2 x_1 + \omega^2 x_1 = \exp{(i\omega T_0)}[-i2D_1 A\omega + i2D_1 \bar{A}\omega\exp{(-i2\omega T_0)}$$

$$- i2\mu A\omega^2 + i2\mu\bar{A}\omega^2 \exp{(-i2\omega T_0)}$$

$$-(A + \bar{A}\exp{(-i2\omega T_0)}) \sum_{r=-n}^{r=n} \cos\left[(1 + r\alpha)T_0 + \phi_r\right]]. \tag{2.145}$$

In order to avoid the secular term problem, it is necessary to state the following

$$D_1 A + \mu \omega A = 0. \qquad (2.146)$$

The particular solution to equation (2.145) can now be ascertained (having proposed equation (2.146)), thus

$$x_1 = \tfrac{1}{2} \sum_{r=-n}^{r=n} \frac{A \exp i[1 + r\alpha + \omega)T_0 + \phi_r]}{[(1 + r\alpha)^2 + 2\omega(1 + r\alpha)]}$$

$$+ \frac{A \exp i[(-1 - r\alpha + \omega) T_0 - \phi_r]}{[(1 + r\alpha)^2 - 2\omega(1 + r\alpha)]} + \text{C.C.} \qquad (2.147)$$

As solutions for x_0 and x_1 have now been found they can be substituted into the second-order perturbation equation (2.143), leading to

$$D_0^2 x_2 + \omega^2 x_2 = -D_0 D_1 \sum_{r=-n}^{r=n} [A\Gamma_1 + A\Gamma_2 + \text{C.C.}] - D_1^2 [A \exp(i\omega T_0)$$

$$+ \bar{A} \exp(-i\omega T_0)] - 2D_2 [i\omega A \exp(i\omega T_0)$$

$$- i\omega A \exp(-i\omega T_0)]$$

$$- 2 \mu \omega D_1 [A \exp(-i\omega T_0) + \bar{A} \exp(-i\omega T_0)]$$

$$- \mu \omega D_0 \sum_{r=-n}^{r=n} [A\Gamma_1 + A\Gamma_2 + \text{C.C}]$$

$$- \tfrac{1}{4} \sum_{r=-n}^{r=n} [A\Gamma_1 + A\Gamma_2 + \text{C.C.}]$$

$$\times \sum_{s=-n}^{s=n} [\exp i[(1 + s\alpha)T_0 + \Phi_s] + \exp i[(-1-s\alpha)T_0 - \phi_s]],$$

$$(2.148)$$

where

$$\Gamma_1 = \frac{\exp i[(1 + r\alpha + \omega)T_0 + \phi_r]}{[(1 + r\alpha)^2 + 2\omega(1 + r\alpha)]}; \qquad (2.149)$$

$$\Gamma_2 = \frac{\exp i[(-1-r\alpha + \omega)T_0 - \phi_r]}{[(1 + r\alpha)^2 - 2\omega(1 + r\alpha)]}. \qquad (2.150)$$

We can express the nearness of the difference in excitation frequencies to twice the natural frequency 2ω by introducing a detuning parameter σ, therefore

$$|r_1 - r_2| \, \alpha = 2\omega + \varepsilon^2 \sigma_m; \quad m = |r_1 - r_2| = 1, 2, 3, \ldots, 2n. \qquad (2.151)$$

Equation (2.151) is substituted into equation (2.150), and the secular term problem is avoided by the removal and setting to zero of the following

$$D_1^2 A + i2\omega D_2 A + 2\mu\omega D_1 A + AP_n + \bar{A}P_{nm}\exp i(\varepsilon^2\sigma_m T_0) = 0, \,(2.152)$$

where

$$P_n = \frac{1}{4}\sum_{r=-n}^{r=n}\left(\frac{1}{(1+r\alpha)^2 - 2\omega(1+r\alpha)} + \frac{1}{(1+r\alpha)^2 + 2\omega(1+r\omega)}\right)$$

$$(2.153)$$

and

$$P_{nm} = \frac{1}{4}\sum_{s=-n}^{s=n}\sum_{r=-n}^{r=n}\left(\frac{\exp i(\phi_s - \phi_r)}{(1+r\alpha)^2 + 2\omega(1+r\alpha)}\right)_{m=s-r}$$

$$+ \sum_{s=-n}^{s=n}\sum_{r=-n}^{r=n}\left(\frac{\exp i(\phi_r - \phi_s)}{(1+r\alpha)^2 - 2\omega(1+r\alpha)}\right)_{m=r-s} \quad (2.154)$$

Because of the algebraic complexity of second order expansions much of the detailed manipulative work cannot be included here. For a more exhaustive treatment the reader is referred to the original sources of this work (Othman, 1986; Othman, Watt and Barr, 1987).

Equation (2.152) is comparable with equation (2.75) but requires rather more lengthy consideration due to the presence of partial derivatives with respect to both time-scales T_1 and T_2.

We can make use of the conditions for secular term avoidance to first order, as given by equation (2.146), and substitute for this as called for in equation (2.152). We also invoke expansions (2.139) and (2.140) so that the ensuring equation can be expressed in terms of real time t. And so this leads to

$$i2\omega\frac{dA}{dt} - \xi^2\omega^2 A + i2\xi\omega^2 A + \varepsilon^2 AP_n + \varepsilon^2\bar{A}P_{nm}\exp i(\varepsilon^2\sigma_m) = 0.$$

$$(2.155)$$

We note that $\xi = \varepsilon\mu$ is the damping ratio (where μ was the ordered form). A solution to equation (2.155) is proposed of the form

$$A = B\exp i[\tfrac{1}{2}\varepsilon^2\sigma_m t], \quad (2.156)$$

where B is a complex amplitude such that

$$B = B_r + iB_i. \quad (2.157)$$

The $\frac{1}{2}$ in the exponent is included to facilitate the cancellation of all exponents in the subsequent equations which result from the substitution of equations (2.156) and (2.157) in (2.155). Therefore two equations come out of this (having separated and grouped real and imaginary terms together), giving

$$\text{Re: } -2\omega \frac{dB_i}{dt} + (\varepsilon^2 C_{1m} + \varepsilon^2 C_{2m} - \omega\varepsilon^2 \sigma_m - \xi^2\omega^2 + \varepsilon^2 P_n)B_r$$

$$+ (\varepsilon^2 S_{1m} + \varepsilon^2 S_{2m} - 2\xi\omega^2)B_i = 0, \qquad (2.158)$$

$$\text{Im: } 2\omega \frac{dB_r}{dt} + (-\varepsilon^2 C_{1m} - \varepsilon^2 C_{2m} - \omega\varepsilon^2 \sigma_m - \xi^2\omega^2 + \varepsilon^2 P_n)B_r$$

$$+ (\varepsilon^2 S_{1m} + \varepsilon^2 S_{2m} + 2\xi\omega^2)B_r = 0, \qquad (2.159)$$

where

$$C_{1m} = \tfrac{1}{4}\sum_{s=-n}^{s=n}\sum_{r=-n}^{r=n} \frac{\cos(\phi_s - \phi_r)}{(1 + r\alpha)^2 + 2\omega(1 + r\alpha)_{m=s-r}}, \qquad (2.160)$$

$$C_{2m} = \tfrac{1}{4}\sum_{s=-n}^{s=n}\sum_{r=-n}^{r=n} \frac{\cos(\phi_r - \phi_s)}{(1 + r\alpha)^2 - 2\omega(1 + r\alpha)_{m=r-s}}, \qquad (2.161)$$

$$S_{1m} = \tfrac{1}{4}\sum_{s=-n}^{s=n}\sum_{r=-n}^{r=n} \frac{\sin(\phi_s - \phi_r)}{(1 + r\alpha)^2 + 2\omega(1 + r\alpha)_{m=s-r}}, \qquad (2.162)$$

$$S_{2m} = \tfrac{1}{4}\sum_{s=-n}^{s=n}\sum_{r=-n}^{r=n} \frac{\sin(\phi_r - \phi_s)}{(1 + r\alpha)^2 - 2\omega(1 + r\alpha)_{m=r-s}}, \qquad (2.163)$$

Equations (2.158) and (2.159) are first-order coupled linear differential equations for which a solution can, in turn, be written, therefore,

$$(B_r, B_i) = (b_r, b_i)\exp(\gamma t). \qquad (2.164)$$

To derive the stability boundary, solution (2.164) is substituted into equations (2.158) and (2.159), and, in order to guarantee non-zero b_r, b_i, the ensuing determinant (constructed from equations (2.158) and (2.159)) is equated to zero. Although the work described within the last few lines is tedious to reproduce it is not conceptually difficult and can therefore be safely omitted. So, we arrive at the point where it is possible to solve for σ_m

$$\sigma_m = \frac{1}{\varepsilon^2\omega}\,[\varepsilon^2 P_n - \xi^2\omega^2$$

$$= \sqrt{\{\varepsilon^4\{(C_{1m} + C_{2m})^2 + (S_{1m} + S_{2m})^2\} - 4\xi^2\omega^2\}]}.\,(2.165)$$

This can be substituted into equation (2.151), and the stability boundary computed. Complementary previous work (Barr and McWhannell, 1971) explored a similar problem in which damping was left out, the analysis used Struble's asymptotic approximations method to treat the equation of motion. The case where several excitation frequencies are nearly coincident was also investigated (Watt and Barr, 1983) and this work is of particular importance because such an excitational function may be considered to be essentially random in form (Rice noise) and therefore redolent of 'natural' excitations on structures. The same single degree of freedom problem was considered in this study under principal parametric resonance, but using the method of Struble and the monodromy matrix technique, a rather complicated method outside the scope of this book.

Results for a three frequency input (with zero damping), where the spacing is $\alpha = 0.1$ (and $\phi = 0°$), such that,

$$f(t) = \cos 0.9t + \cos t + \cos 1.1t$$

are given in Fig. 2.9. The boundaries are given as functions of the $1/2\omega - \varepsilon$ plane (where $1/2\omega = 1/(r_1 - r_2)\alpha = 1/(m\alpha)$). The method is seen to hold quite well for small ε until $\varepsilon \simeq 0.04$ at which point the second zone (coming out from $1/2\omega = 5$ – for which $m = 2$, $\alpha = 0.1$) overlaps the first ($1/2\omega = 10$; $m = 1$, $\alpha = 0.1$). Overlaps would not occur in practice, and the analysis is weak at this point and for increasing ε. In the original reference (Othman, 1986; Othman, Watt and Barr, 1987) there is comparative data based on Stuble's method and the monodromy matrix method. Boundaries for the damped case lie entirely within the undamped stability boundaries since the rates of solution growth are identical in the two cases, and as one would expect the damped case does not commence right at the $1/2\omega$ axis but is pushed to higher ε values dependent on the operative damping level itself.

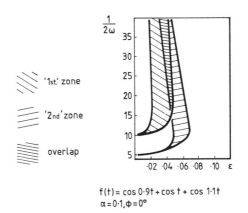

$$f(t) = \cos 0.9t + \cos t + \cos 1.1t$$
$$\alpha = 0.1, \phi = 0°$$

Figure 2.9 Stability chart showing the transition curve boundaries for a three frequency input excitation (principal parameteric resonance of the form of equation (2.151)) (after Othman, 1986 and Othman, Watt and Barr, 1987).

(b) Simultaneous forced and parametric excitation

This is a complex problem if complete independence between the external and parametric resonances is sought. We initially examine one simple case here for which the parametric excitation frequency, Ω_2, is fixed at twice the external force frequency Ω_1, so that detuning of one frequency automatically decides that of the other.

Therefore,

$$\Omega_1 = \omega + \varepsilon\gamma \tag{2.166}$$

and

$$\Omega_2 = 2\omega + \varepsilon\delta. \tag{2.167}$$

Taking the condition

$$\Omega_2 = 2\Omega_1 \tag{2.168}$$

we get

$$\delta = 2\gamma. \tag{2.169}$$

The equation of motion is taken to be

$$\ddot{x} + 2\xi\omega\dot{x} + \omega^2 x - E_P x \cos\Omega_2 t = E_F \cos\Omega_1 t. \tag{2.170}$$

E_P and E_F represent parametric and forcing excitation magnitudes and, together with damping, are expressed to first-order ε, therefore,

$$E_P = \varepsilon e_P; \qquad E_F = \varepsilon e_F; \qquad \xi = \varepsilon\zeta.$$

Expansion of the variable and its derivatives up to and including first-order ε leads to

$$\varepsilon^0: D_0^2 x_0 + \omega^2 x_0 = 0, \tag{2.171}$$

$$\varepsilon^1: D_0^2 x_1 + \omega^2 x_1 = -2D_0 D_1 x_0 - 2\zeta\omega D_0 x_0$$
$$+ x_0 e_P \cos\Omega_2 t + e_F \cos\Omega_1 t. \tag{2.172}$$

Taking the polar form for the solution of equation (2.171)

$$x_0 = A \exp i\omega T_0 + A \exp i(-\omega)T_0 \tag{2.173}$$

and substituting for this in equation (2.172) leads to

$$D_0^2 x_1 + \omega^2 x_1$$
$$= \exp i\omega T_0 [-i2D_1 A\omega + i2D_1 \bar{A}\omega \exp i(-2\omega T_0) - i2\zeta\omega^2 A$$
$$+ i2\zeta\omega^2 \bar{A} \exp i(-2\omega T_0)$$
$$+ \frac{e_P}{2} (A \exp i\Omega_2 T_0 + A \exp i(-\Omega_2)T_0$$
$$+ \bar{A} \exp i(\Omega_2 - 2\omega)T_0 + \bar{A} \exp i(-\Omega_2 - 2\omega)T_0)$$

$$+ \frac{e_F}{2} (\exp i(\Omega_1 - \omega)T_0 + \exp i(-\Omega_1 - \omega)T_0]. \quad (2.174)$$

Substituting equations (2.166) and (2.167) into the equation (2.174) and removing troublesome terms (those that will generate secular terms) and equating to zero leaves

$$-i2D_1 A\omega - i2\zeta\omega^2 A + \frac{e_P}{2} \bar{A} \exp i\varepsilon\delta T_0 + (e_F/2) \exp i\varepsilon\gamma T_0 = 0. \quad (2.175)$$

A solution for A is of the following form

$$A = \frac{a}{2} e^{i\alpha}; \quad \bar{A} = \frac{a}{2} e^{-i\alpha}, \quad (2.176)$$

which, when substituted into equation (2.175) yields

$$-i\omega a' + \omega a\alpha' - i\zeta\omega^2 a + \frac{e_P}{4} a \exp i(\varepsilon\delta T_0 - 2\alpha)$$

$$+ \frac{e_F}{4} \exp i(\varepsilon\gamma T_0 - \alpha) = 0. \quad (2.177)$$

Separation into the real and imaginary parts gives

$$\text{Re: } \omega a\alpha' + \frac{e_P}{4} a \cos(\varepsilon\delta T_0 - 2\alpha) + \frac{e_F}{4} \cos(\varepsilon\gamma T_0 - \alpha) = 0, \quad (2.178)$$

$$\text{Im: } -\omega a' - \zeta\omega^2 a + \frac{e_P}{4} a \sin(\varepsilon\delta T_0 - 2\alpha) + \frac{e_F}{4} \sin(\varepsilon\gamma T_0 - \alpha) = 0. \quad (2.179)$$

We make the system autonomous by defining the trigonometrical function arguments as

$$\Phi = \varepsilon\delta T_0 - 2\alpha, \quad (2.180)$$

$$\Psi = \varepsilon\gamma T_0 - \alpha. \quad (2.181)$$

Substituting for δ (from equation (2.169)) into equation (2.180) gives

$$\Phi = 2(\varepsilon\gamma T - \alpha) \quad (2.182)$$

and then substitution of equation (2.181) into (2.182) leads to

$$\Phi = 2\Psi. \quad (2.183)$$

This is of course only true if equation (2.168) holds.

Differentiating equation (2.181) with respect to time-scale T_1, noting that

$$\Psi = \varepsilon\gamma T_0 - \alpha = \gamma T_1 - \alpha. \quad (2.184)$$

Therefore,

$$\Psi' = \gamma - \alpha'. \quad (2.185)$$

And so we have

$$\alpha' = \gamma - \Psi'. \tag{2.186}$$

We can now rewrite equations (2.178) and (2.179) making the necessary substitutions

$$\omega a(\gamma - \Psi') + \frac{e_P}{4} a \cos 2\Psi + \frac{e_F}{4} \cos \Psi = 0, \tag{2.187}$$

$$-\omega a' - \zeta \omega^2 a + \frac{e_P}{4} a \sin 2\Psi + \frac{e_F}{4} \sin \Psi = 0. \tag{2.188}$$

Multiplying both equations through by ε allows a return to the original parameters of the differential equation

$$\dot{\Psi} = \varepsilon \gamma + \frac{E_P}{4\omega} \cos 2\Psi + \frac{F_\Gamma}{4\omega a} \cos \Psi, \tag{2.189}$$

$$\dot{a} = -\xi \omega a + \frac{E_P}{4\omega} a \sin 2\Psi + \frac{E_F}{4\omega} \sin \Psi, \tag{2.190}$$

where the dot denotes differentiation with respect to time-scale T_0 ($T_0 \simeq t$).

Equations (2.189) and (2.190) may be integrated for chosen numerical data, and a selection of results are given in Fig. 2.10(a) to 2.10(f). It should be noted that the foregoing analysis is restricted solely to the exact condition of equation (2.168) and that the general case

$$\Omega_2 = 2\Omega_1 + \varepsilon \beta \tag{2.191}$$

requires a little bit of further analysis.

In order to deal with this we substitute equations (2.166) and (2.167) into (2.191). This leads to

$$\delta = 2\gamma + \beta. \tag{2.192}$$

Substituting for δ in equation (2.180) gives a redefined Φ

$$\Phi = 2\Psi + \varepsilon \beta T_0 \tag{2.193}$$

enabling us to rewrite equations (2.189) and (2.190), thus,

$$\dot{\Psi} = \varepsilon \gamma + \frac{E_P}{4\omega} \cos (2\Psi + \varepsilon \beta T_0) + \frac{E_F}{\omega a} \cos \Psi, \tag{2.194}$$

$$\dot{a} = -\xi \omega a + \frac{E_P}{4\omega} a \sin (2\Psi + \varepsilon \beta T_0) + \frac{E_F}{4\omega} \sin \Psi. \tag{2.195}$$

Introducing a further autonomous angle Δ, where,

$$\Delta = 2\Psi = \varepsilon \beta T_0. \tag{2.196}$$

Therefore,

$$\dot{\Delta} = 2\dot{\Psi} + \varepsilon \beta. \tag{2.197}$$

So finally we have a set of three first-order equations which describe the general system based on resonance conditions (2.166), (2.167), (2.191),

$$\dot{a} = -\xi\omega a + \frac{E_P}{4\omega} a \sin \Delta + \frac{E_F}{4\omega} \sin \Psi, \tag{2.198}$$

$$\dot{\Psi} = \varepsilon\gamma + \frac{E_P}{4\omega} \cos \Delta + \frac{E_F}{4\omega a} \cos \Psi, \tag{2.199}$$

$$\dot{\Delta} = 2\dot{\Psi} + \varepsilon\beta. \tag{2.200}$$

These equations can be integrated numerically, however no results are presented here since a full treatment of this class of problem is currently under investigation (Watt and Santos, University of Aberdeen) and will be reported in the literature in due course.

For the moment we can examine some of the behaviour of the specific problem where the parametric excitation frequency is fixed at twice the excitation frequency.

Figure 2.10(a) shows the case where the parametric excitation is zero and the forcing level is fixed at three points, $E_P = 10$, 50 and 100. As one would expect response a settles down into steady state well before the end of the integration period. The detuning is zero here, $\varepsilon\gamma = \varepsilon\delta = 0$ and so these curves represent the peaks of the linear forced vibration responses that could be generated for each forcing level. Variation of $\varepsilon\gamma$ around zero would enable such frequency response plots to be constructed. If we now put the forcing term to zero and apply a parametric excitation we would expect conventional parametric behaviour as discussed in Section 2.3.3, particularly since the detuning is still zero ($\varepsilon\gamma = \varepsilon\delta = 0$) thus guaranteeing that we are considering a point (centrally) within the unstable zone. This is indeed so as long as the parametric excitation is above the threshold level. The threshold is a function of damping, the higher the damping the higher the excitation threshold. For the case of the numerical data used the threshold is at $E_P = 40$, and so we see from Fig. 2.10(b) that $E_P = 39$ generates a response that decays to zero from the initial condition ($a(0) \neq 0$ because of the form of equation (2.189)). $E_P = 41$ shows a climb which is substantiated by further integration over a longer time period; therefore provided $E_P > 40$ we will find that $a \to \infty$, the rate of growth depending on the actual value of E_P. Figure 2.10(c) illustrates several simultaneous forced and parametric cases all for exact tuning $\varepsilon\gamma = \varepsilon\delta = 0$, and it is apparent that for some fixed forcing level ($E_F = 50$ was chosen here) the parametric excitation serves to increase the eventual steady state response amplitude as long as $E_P < 40$. So subthreshold parametric excitations, which would not initiate any response at all on their own exaggerate the effect of the forcing. The case where $E_P > 40$ for

$E_F = 50$ is however different and Fig. 2.10(d) verifies that eventually a becomes unbounded. Reference to Fig. 2.10(e) shows how the case of $E_P = 39$, $E_P = 50$ conforms to the steady state behaviour previously

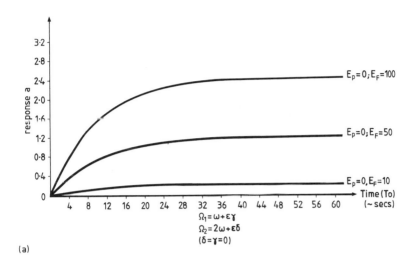

(a)

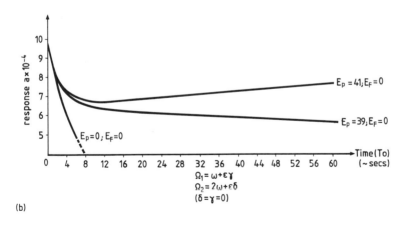

(b)

Figure 2.10 Responses of a theoretical system under combined forced and parametric excitations: (a) case for parametric excitation zero $(E_p = 0)$; (b) threshold parametric excitation case, zero forced excitation $(E_p \neq 0; E_F = 0)$; (c) various cases; (d) post-threshold parametric excitation with medium forcing; (e) subthreshold parametric excitation with medium forcing; (f) various cases, all for detuned parametric and forced excitations.

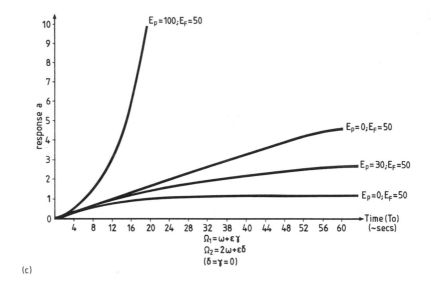

(c)

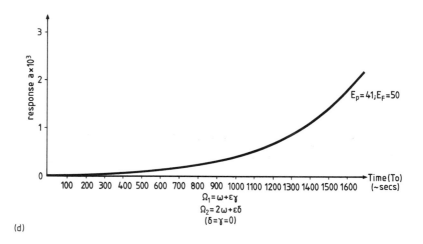

(d)

described, however it takes a long time to reach this point. The three remaining cases in Fig. 2.10(c) show how different excitation levels interact with a strong parametric excitation of $E_P = 100$. As E_F is increased from 0 to 100 the rate of growth of the unbounded response with time increases. These cases are also true for the perfectly tuned case, $\varepsilon\gamma = \varepsilon\delta = 0$. Finally the condition for which $\varepsilon\gamma \neq 0$; $\varepsilon\delta \neq 0$ is cursorily presented in Fig. 2.10(f) for three pairs of excitation levels.

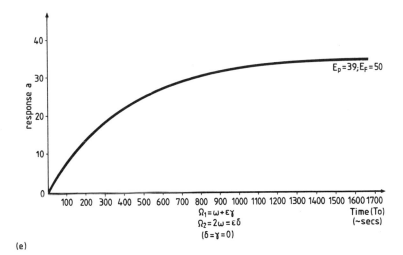

(e)

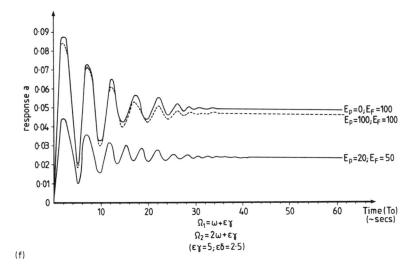

(f)

The detuning is now $\varepsilon\gamma = 5$; $\varepsilon\delta = 2.5$ and we can interpret this as meaning that Ω_1 is 5% of ω away from exact external tuning (i.e. above it) and that Ω_2 is 2.5% of ω above exact principal parametric resonance. Therefore in the forced only case the response would be off resonance, and in the parametric only case it would be away from the centre of the unstable zone (whether it is inside this or not depends on the magnitudes E_P and ξ).

The time response of α for $E_P = 20$; $E_F = 50$; $\varepsilon\gamma = 5$; $\varepsilon\delta = 2.5$ shows decaying oscillatory behaviour for two thirds of the integration time which eventually settles out to steady state. This curve is directly comparable with the exactly tuned case of Fig. 2.10(c) and shows a very reduced eventual steady-state response amplitude with detuning in force. Doubling the forcing level doubles the eventual response (case of $E_P = 0$; $E_F = 100$; $\varepsilon\gamma = 5$; $\varepsilon\delta = 2.5$ in Fig. 2.10(f)) with the reduced parametric excitation having little or no effect. Interestingly enough increasing the parametric excitation to $E_P = 100$ for the case of $E_F = 100$ merely serves to reduce the steady state response slightly. It is clear that for $0 \leqslant E_P \leqslant 100$ the case $\varepsilon\delta = 2.5$ takes the system outside the unstable zone.

(c) Higher-order combination resonances

As a rule these are generated through coupling either in the form of apparently small non-linearities or coupling due to parametric excitation terms in the equations. Higher-order resonances are generally reckoned to be weaker than those predicted by first-order theory and so in practice necessitate higher excitation levels to initiate them, also the amount of analytical work required to uncover the form of a higher-order resonance is very much greater than that needed for first-order effects; compare the examples of Section 2.3.1 and Section 2.4.1 (first example). This does not mean that first-order resonances cannot be treated to second order, however the additional work would generally not justify the modest increase in accuracy obtained and therefore such a course would normally be avoided. We consider the following system which has its physical basis in Fig. 2.5(d).

$$\ddot{x}_1 + 2\xi_1\omega_1\dot{x}_1 + \omega_1^2 x_1 - F_1 \cos \Omega t . x_3 = 0, \tag{2.201}$$

$$\ddot{x}_2 + 2\xi_2\omega_2\dot{x}_2 + \omega_2^2 x_2 - F_2 \cos \Omega t . x_3 = 0, \tag{2.202}$$

$$\ddot{x}_3 + 2\xi_3\omega_3\dot{x}_3 + \omega_3^2 x_3 - F_{31} \cos \Omega t . x_1 - F_{32} \cos \Omega t x_2 = 0. \tag{2.203}$$

The excitation is redefined to first-order ε and the damping terms (which are likely to be small) are put to second-order ε, therefore,

$$F_1 = \varepsilon f_1; \qquad F_2 = \varepsilon f_2; \qquad F_{31} = \varepsilon f_{31}; \qquad F_{32} = \varepsilon f_{32};$$

$$\xi_k = \varepsilon^2 \zeta_k, \qquad k = 1, 2, 3.$$

Equations (2.201) to (2.203) are treated by means of a second-order multiple scales expansion with substitutions in the form of equations (2.138)–(2.140) in order to generate the zeroth-, first- and second-order perturbation equations. The problem was originally worked as one containing a large number of cubic non-linearities (Cartmell and

Roberts, 1987) and for this reason a full review of the analysis will be held over until Chapter 3. As will be apparent equations (3.66)–(3.68) have been reduced to the above set (2.201)–(2.203) since it is these retained terms that can be shown to be the source of one particularly interesting combination

$$\Omega = \tfrac{1}{2}(\omega_1 + \omega_2) + \varepsilon^2 \rho_1. \tag{2.204}$$

An expression for the detuning parameter $\varepsilon^2 \rho_1$ can be obtained by determining a set of autonomous slow time equations (with respect to time-scale T_2) and then by setting the slowly varying amplitudes and autonomous phase angles to zero. The slow time equations are derived from the right-hand sides of the second-order perturbation equations (3.81)–(3.83) in the usual manner. The result for $\varepsilon^2 \rho$, is a quadratic,

$$V_1(\varepsilon \rho_1^2)^2 + V_2(\varepsilon \rho_1^2) + V_3 = 0, \tag{2.205}$$

where the V_i are rather complicated coefficients involving system constants and the excitation function; refer to Chapter 3 for further details. Figure 2.11 contains the transition curves for resonance (2.204) (two beam lengths), and also the first-order effect defined by equation (2.74). Confirmatory experimental points for the stability boundary are also given here. The unstable zone of the second-order resonance for the shorter beam is altogether smaller than that of the first-order effect and well back from the frequency axis, but comes nearer with increasing beam length. For the laboratory model it is found to be relatively easy to get into the unstable zone, and once inside this the response begins to grow with time in an unbounded manner. Integration of the slow time equations enables this growth to be predicted theoretically, this is shown in Fig. 2.12.

If we reconsider the kinematics of the structure of Fig. 2.5(d) then it becomes apparent that any response that is out of the plane of excitation must involve both bending and torsional motion. Therefore the system kinematics readily admit resonance (2.74), but it is less obvious how resonance (2.204), which does not involve torsion at all, might come about. It is found from spectral analysis of the beam's response before, during and after the instability develops that in the early stages there is a certain amount of torsional motion present, and all three modes are responding and exchanging energy between themselves. Eventually the torsional response falls away and the two bending modes respond alone. In practice, response growth is limited by certain stabilizing aspects of the system and very large, but steady, modal vibrations ensue. This becomes steady-state behaviour due to the influence of nonlinear effects stabilizing the basically unstable linear problem.

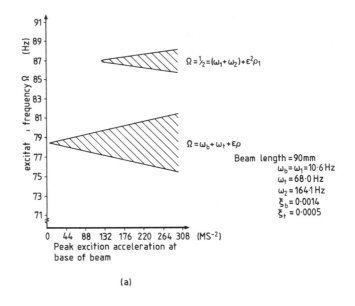

(a)

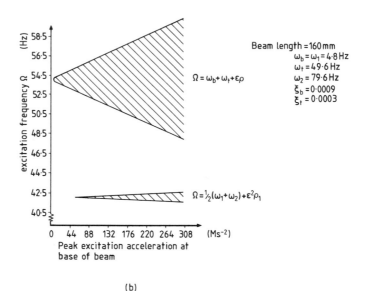

(b)

Figure 2.11 Stability chart showing a second-order bending/bending combination transition curve and the first-order bending/torsion combination transition curve; two cases (a) and (b), both for a cantilevered beam.

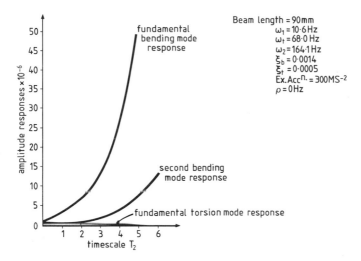

Figure 2.12 Theoretical responses of second-order bending/bending instability in a cantilever beam (as functions of slow time-scale T_2).

The mechanism behind the early torsional presence (which seems to provoke the eventual bending instability of equation (2.204)) may have more than one source but it is considered likely that a slightly offset mass centre together with a fairly high excitation level could be sufficient to generate a small torsional component which thus provides the requisite kinematical coupling. It should be noted that the longer (and therefore the more flexible) the beam the more the unstable zone of resonance (2.204) approaches the frequency axis (and the narrower it gets), and so the lower the excitation level required in order to generate it.

(d) Parametric excitation with a fluctuating frequency

Additional to the work of Othman, Watt and Barr is a recently reported investigation into parametric excitation of a linear, single degree of freedom oscillator where the parametric excitation has a fluctuating frequency (Miyasar and Barr, 1988); this is equivalent to the multi-frequency input problem of Watt and Barr (Watt and Barr, 1983), refer to Section 2.4.1(a). Although these two problems are quite different they have some commonality in that they present possibilities which are not covered by the single, fixed frequency, sinusoidal excitation case. Miyasar and Barr propose a system governed by the following

$$\ddot{Y} + 2\xi(\Omega + d)\dot{Y} + \{(\Omega + d)^2 + \varepsilon\cos\mu(t)\}y = 0 \qquad (2.206)$$

for which the parametric excitation function is $\varepsilon \cos(2\Omega t + \alpha \cos \beta t)$. The excitation frequency varies around a mean value 2Ω sweeping through β with amplitude $\alpha\beta$, and $(\Omega + d)$ represents the undamped system natural frequency with d defining some degree of detuning. The problem is treated by means of Struble's method to give a pair of nonlinear coupled first-order equations for response amplitude and phase. It is likely that similar if not identical expressions would result from the use of another method of analysis, such as multiple scales. The first-order set is

$$\dot{\Phi} = -\alpha\beta \sin \beta t - \frac{\varepsilon}{2(\Omega + d)} \cos \Phi - 2d, \qquad (2.207)$$

$$\dot{A} = -A \frac{\varepsilon}{4(\Omega + d)} \sin \Phi - \xi(\Omega + d), \qquad (2.208)$$

where Φ and A are the phase and amplitude respectively.

Numerical integration of equations (2.207) and (2.208) enables us to propose the following general statements about the behaviour of the system.

For $\alpha = 0$, where we have the straightforward Mathieu equation, the phase angle Φ grows until it reaches a point at which it remains fairly steady for increasing t. The amplitude will increase monotonically which means that it is strictly increasing or decreasing over an interval of time. By making $\alpha > 0$ the amplitude growth rate tends to decrease, reaching a minimum at $\alpha \simeq 2.5$, but after this it increases again. The phase oscillates about a mean such that the overall magnitude of oscillation is 2α.

The effect of increasing the sweep frequency β is to increase the oscillation frequency of Φ and to slightly reduce the amplitude growth rate. Increasing the value of the excitation parameter ε provides a corresponding increase in the rate of amplitude growth. As is always the case in parametric vibrations problems the influence of detuning about resonance tends to be significant. This case is no exception. Increasing d will reduce the rate of growth of the response (except for very small values) until the solution passes out of the defined instability region. Eventually, if d is increased sufficiently, the system would be likely to enter further instability zones; these zones are all shown to be to order ε. The phase angle will grow with d, the actual extent of the phase magnitude being dependent on α.

Damping has considerable effect upon the response amplitude stability but no significant phase variation occurs. Thus increasing ξ decreases the extent of the instability boundaries. For complete coverage of the problem including several stability boundaries the reader is recommended to read this work in full (Miyasar and Barr, 1988).

2.4.2 Machine problems, power transmission chains

(a) Moving chain problem

One of the earliest studies in parametric vibrations of travelling power transmission devices (Mahalingam, 1957) showed that consideration of the chain as a heavy uniform string undergoing a longitudinal excitation admitted the possiblity of a lateral, parametric, instability. The problem is shown diagrammatically in Fig. 2.13. Taking an $X–Y$ coordinate

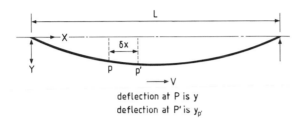

deflection at P is y
deflection at P' is $y_{p'}$

Figure 2.13 Lateral deflection schematic for moving chain problem (after Mahalingam, 1957).

system for the chain and defining its longitudinal velocity by V we see that initially the deflection of some arbitrary point P along the unsupported length is y, and that this will change as the chain moves to another point, P', through time δt. The lateral deflection will now be given by

$$y_{p'} = y + \frac{\partial y}{\partial x}\,\delta x + \frac{\partial y}{\partial t}\,\delta t. \tag{2.209}$$

But since

$$V = \frac{\delta x}{\delta t} \tag{2.210}$$

equation (2.209) becomes

$$y_{p'} = y + \left(V\frac{\partial}{\partial x} + \frac{\partial}{\partial t}\right)y\,\delta t. \tag{2.211}$$

Therefore the transverse velocity of P' is

$$V_{p'(\text{trans})} = \left(V\frac{\partial}{\partial x} + \frac{\partial}{\partial t}\right)y. \tag{2.212}$$

Thus the tranverse acceleration of P' is

$$a_{p'(\text{trans})} = \left(V\frac{\partial}{\partial x} + \frac{\partial}{\partial t}\right)^2 y. \tag{2.213}$$

The equation of motion is

$$T \frac{\partial^2 y}{\partial x^2} = m a_{p'(\text{trans})}, \qquad (2.214)$$

where T is the chain tension.

Thus we have

$$\left(\frac{T}{m}\right) \frac{\partial^2 y}{\partial x^2} = V^2 \frac{\partial^2 y}{\partial x^2} + 2V \frac{\partial^2 y}{\partial x \partial t} + \frac{\partial^2 y}{\partial t^2}. \qquad (2.215)$$

(T/m) may be written as V_0^2, since V_0 may be defined as the velocity of transverse wave propagation, therefore we now have

$$V_0^2 \frac{\partial^2 y}{\partial x^2} = V^2 \frac{\partial^2 y}{\partial x^2} + 2V \frac{\partial^2 y}{\partial x \partial t} + \frac{\partial^2 y}{\partial t^2}. \qquad (2.216)$$

A longitudinal excitation may be introduced in the form of a torsional oscillation of one of the sprockets. This can be expressed as $a \cos \omega t$ where ω is the frequency of torsional oscillation of the sprocket. The quantity a represents the periodic length change that the chain undergoes and so we can represent the instantaneous tension of the chain as $T + \Delta T \cos \omega t$. Substitution of this into equation (2.215) via our definition of V_0 leads to

$$(V_0^2 + \Delta V_0^2 \cos \omega t) \frac{\partial^2 y}{\partial x^2} = V^2 \frac{\partial^2 y}{\partial x^2} + 2V \frac{\partial^2 y}{\partial x \partial t} + \frac{\partial^2 y}{\partial t^2}. \qquad (2.217)$$

Because $y = y(x, t)$ we require a transformation into a coordinate with dependency solely on x or t. We choose to transform into $y_0 = y_0(t)$, where

$$y(x, y) = y_0(t)(\cos nx + i \sin nx) = y_0(t) e^{inx} \qquad (2.218)$$

where n can remain unspecified for the present.

Substituting equation (2.218) into (2.217) enables the governing equation (now with time dependency only) to be rewritten

$$-(V_0^2 + \Delta V_0^2 \cos \omega t) y_0 n^2 e^{inx} = -V^2 y_0 n^2 e^{inx} + 2V \dot{y}_0 in e^{inx} + \ddot{y}_0 e^{inx}. \qquad (2.219)$$

Cancelling e^{inx} and rearranging gives

$$\ddot{y}_0 + i2Vn\dot{y}_0 + [n^2(V_0^2 - V^2 + \Delta V_0^2 \cos \omega t)]y_0. \qquad (2.220)$$

Equation (2.220) is beginning to resemble the Mathieu form and in order to complete the analysis we make our further transformation whereby we eliminate the second term of equation (2.220).

We do this by introducing a further variable in such a way that we have

$$y_0(t) = e^{-iVnz} u(z). \qquad (2.221)$$

Argument ωt is restated also

$$\omega t = 2z. \tag{2.222}$$

Differentiation of equation (2.221) with respect to time so that we have

$$\dot{y}_0 = \frac{dy_0}{dz}\frac{dz}{dt}; \qquad \ddot{y}_0 = \frac{d^2 y_0}{dz^2}\left(\frac{dz}{dt}\right)^2,$$

enables the following substitutions to be made in equation (2.220)

$$[-V^2 n^2 u(z) - i2Vnu'(z) + u''(z)]\left(\frac{dz}{dt}\right)^2$$

$$+ i2Vn[-iVnu(z) + u'(z)]\left(\frac{dz}{dt}\right) + [n^2 V_0^2 u(z)$$

$$- n^2 V^2 u(z) + n^2 \Delta V_0^2 \cos 2z . u(z)] = 0, \tag{2.223}$$

where the primes denote differentiation with respect to z.

If we stipulate that $z = t$, then $\omega = 2$, and we can revert to the dot derivative notation; expanding equation (2.223) gives

$$\ddot{u} + (n^2 V_0^2 + n^2 \Delta V_0^2 \cos 2t)u = 0. \tag{2.224}$$

We can simplify further by stating

$$\omega_0^2 = V_0^2; \qquad \varepsilon = \Delta V_0^2, \tag{2.225}$$

which leads to

$$\ddot{u} + (n^2 \omega_0^2 + n^2 \varepsilon \cos 2t)u = 0. \tag{2.226}$$

This is a linear undamped Mathieu–Hill equation similar to equation (2.1) for which we can easily derive the stability boundary for principal parametric resonance by considering equation (2.131) where $\xi = 0$ and $\Omega = \omega = 2$.

Therefore we arrive at the following

$$\frac{1}{n} = \omega_0 \pm \frac{\varepsilon}{4\omega_0}. \tag{2.227}$$

This is shown in Fig. 2.14 (note that $\omega = 2\omega_0$ for this principal parametric resonance case). The excitation is provided by ε and we can relate this back to chain tension since

$$\varepsilon = \frac{\Delta T}{M}. \tag{2.228}$$

Therefore the larger the torsional oscillation of the sprocket the larger the magnitude of ΔT, and hence ε, and the wider the region of instability at principal parametric resonance. Substitution of $\varepsilon = 1, 2, 3,$

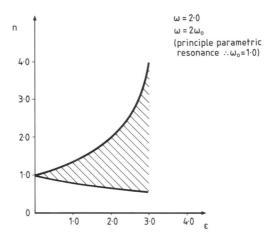

Figure 2.14 Stability chart showing the transition curve and unstable zone for principal parametric resonance in the moving chain problem (data and plot derived by the author with reference to the work of Mahalingam, 1957).

... into equation (2.227) enables the instability region for $\omega = 2\omega_0$; $\omega = 2$ to be portrayed; see Fig. 2.14.

(b) Parametric vibrations in band saw blades and pulley belts

Moving saw blade and pulley belt problems are very similar to the previous chain example in that a tension fluctuation caused by wheel eccentricity or by joints and flaws in the band is the main source of parametric excitation. Experimental work has indicated that irregularities in belt thickness tend to be predominantly responsible for inducing parametric vibrations in vee belt systems.

The analysis of such systems is similar to that given above, an interesting example having been given by Naguleswaran and Williams (1967); see Fig. 2.15. In this work it was observed that although the mean tension of the band or belt increases with speed V the natural frequencies of the various modes of free vibration actually decrease with V. This seems difficult to accept on an intuitive basis, and therefore we need to consider the effect of longitudinal band motion on the speed of wave propagation within the band for an explanation. If $V = 0$ the wave takes $2L/c_0$ units of time to traverse the span L in both directions, where the speed of wave propagation c_0 is given by,

$$c_0 = \sqrt{\frac{T_0}{m}}$$ (2.229)

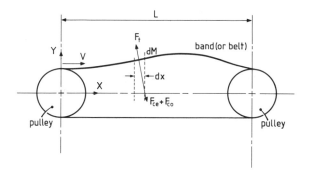

Figure 2.15 Lateral defelection schematic for moving belt or blade problem (after Naguleswaran and Williams, 1967).

T_0 being the static belt tension, and m the mass per unit length of band or belt. Thus the fundamental natural frequency is $c_0/(2L)$. On the other hand for a moving belt $V \neq 0$, and the absolute 'upstream' and 'downstream' propagation velocities become $(c - V)$ and $(c + V)$, taking 'downstream' to denote the direction of V. The moving belt tension T_b is given by

$$T_b = T_0 + \eta m V^2. \tag{2.230}$$

The speed of wave propagation in the moving band is therefore

$$c = \sqrt{\frac{T_b}{m}} \tag{2.231}$$

and is relative to the band. The fundamental periodic time is now $L/(c + v) + L/(c - V)$, or $2cL/(c^2 - V^2)$, and clearly the fundamental frequency will be $(c^2 - V^2)/(2cL)$ which diminishes with increasing V. Combining equations (2.230) and (2.231) and substituting for c in the fundamental frequency expression for the moving band leads to

$$\omega_{fb}^2 = \frac{(T_0/m + V^2[\eta - 1])^2}{4L^2(T_0/m + \eta V^2)}. \tag{2.232}$$

The variation of ω_{fb} with band speed V is given in Fig. 2.16 for $\eta = 0$ and $\eta = 1$, clearly showing the effect of V on ω_{fb} and also the significance of the centrifugal component denoted by η. The extreme cases are where $(\eta = 1)$ (for which T_b is a maximum for some given V; equation (2.20)) and here the supporting structure is stiffest, and, although ω_{fb} still decreases with V, the curve eventually becomes asymptotic. Conversely if $\eta = 0$ (therefore $T_b = T_0$; equation (2.230)) the decrease is more pronounced.

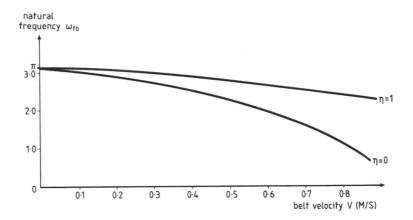

Figure 2.16 Variation of the fundamental natural frequency of lateral vibration of a moving belt as a function of belt velocity; for included centrifugal component ($\eta = 1$) and also for zero centrifugal component ($\eta = 0$) (after Naguleswaran and Williams, 1967).

By considering an element of the band or belt the equation of motion is

$$F_t - F_{ce} - F_{co} - F_i = 0, \tag{2.233}$$

where F_t, F_{ce}, F_{co} and F_i are, respectively, the force on the element due to belt tension, the centrifugal force due to belt curvature, the coriolis force due to combined linear and rotational motion of the element, and the elemental interia force.

Adding a harmonically varying excitation term into equation (2.230) leads to

$$T_b = T_0 + \eta m V^2 + \delta T_0 \cos \Omega T. \tag{2.234}$$

Therefore for the full system we have

$$(T_0 + \eta m V^2 + \delta T_0 \cos \Omega t - m V^2) \frac{\partial^2 Y}{\partial X^2}$$

$$- 2mV \frac{\partial^2 Y}{\partial X \partial T} - m \frac{\partial^2 Y}{\partial T^2} = 0. \tag{2.235}$$

This bears close resemblance to the equation for the chain (equation (2.215)); the difference really being in the centrifugal aspect to the band tension term. In the original paper the authors develop equation (2.235) a little further by non-dimensionalizing it such that,

$$(1 + \varepsilon \cos \omega t - k v^2) \frac{\partial^2 y}{\partial x^2} - 2v \frac{\partial^2 y}{\partial x \partial t} - \frac{\partial^2 y}{\partial t^2} = 0, \tag{2.236}$$

where $X = xL$; $Y = yL$; $T = (tL \sqrt{m/T_0})$; $\Omega = (\omega \sqrt{T_0/(mL^2)})$; $\varepsilon = (\delta T_0/T_0)$, $k = 1 - \eta$; $v = (V\sqrt{m/T_0})$.

Rearranging this equation allows direct comparison with equation (2.217)

$$(1 + \varepsilon \cos \omega t) \frac{\partial^2 y}{\partial x^2} = kv^2 \frac{\partial^2 y}{\partial x^2} + 2v \frac{\partial^2 y}{\partial x \partial t} + \frac{\partial^2 y}{\partial t^2}. \qquad (2.237)$$

Following the same procedure as in the chain problem we can derive a Mathieu type equation similar in form to equation (2.224)

$$\ddot{u} - \left(n^2 v^2 \left[1 - k + \frac{1}{v^2} \right] + n^2 \, \varepsilon \cos \omega t \right) u = 0. \qquad (2.238)$$

In this case we put

$$\omega_0^2 = v^2 \left[1 - k + \frac{1}{v^2} \right] \qquad (2.239)$$

which results in an equation identical to equation (2.226) except with a redefined natural frequency. Clearly the stability curve for equation (2.227) will also apply here.

The literature dealing with belt and band problems is fairly extensive and could also quite feasibly be considered to cover moving string problems as well. The latter have been investigated under parametric excitation (Mote, 1968), and regions of parametric instability determined for various string 'transport' velocities.

(c) *Brief review of a parametric self-excitation problem*

Parametric self-excitation of a belt into tranverse vibration has been explored for the case of a belt and two pulleys (Rhodes, 1970), and this is a particularly interesting variation on the previous themes of imperfect pulley/gear mountings or flawed belts. It demonstrates that natural variations in belt tension, on take up round a pulley, are 'recorded' as a stress pattern within the belt (provided there is no slip) and that this stress pattern, on pay off on to the next free span, reproduces the tension variation into this portion of belt. The belt passes around the other pulley, and thus transmits this travelling tension variation round the system so that there is a delay between the originally generated variation and its re-emergence at the same physical point. The author points out that if this delay is closely related to the period of transverse vibration of the belt a parametric vibration can result which under certain circumstances will increase in magnitude (such circumstances might, for example, include the case of low frequency modes for which the attenuation of the circulated tension variation is minimized). Any imperfection within the belt or pulley would be likely to exacerbate the situation even more.

Several parametrically resonant effects have been discovered in an axially moving band with an in-plane periodic edge loading normal to the longitudinal axis of the band (Wu and Mote, 1986). Here the edge loading could be generated by the sort of cutting forces that are exerted on a moving band saw blade for example.

2.4.3 Parametric vibrations in engineering structures

Combination resonances due to lateral bending–torsion vibrations of a thin beam under parametric excitation have been considered in earlier sections of this chapter (Sections 2.3.2 and 2.3.3) and an expression for the transition curve derived (Yamamoto and Saito, 1970; Dugundji and Mukhopadhyay, 1973).

(a) Axial static and dynamic loading of a column

We will examine a reported case of structural parametric vibration which broadly relates to mechanical and civil engineering. It is not uncommon to find structural members loaded axially by means of a static dead-load and then excited into vibration due to an additional dynamic loading. The dynamic load could arise because of seismic disturbances or bomb blast effects (in the case of civil structures) or mechanical disturbances of some sort. The force exerted on a column, for example, could therefore be given by

$$P = P_0 + P_1 \sin \Omega_1 t, \tag{2.240}$$

where the static and dynamic aspects of P are given by P_0 and $P_1 \sin \Omega_1 t$ respectively.

We could examine the response of a column to this type of loading in order to predict conditions under which it might respond laterally in the form of a parametric instability. A diagrammatic representation of such a problem is given in Fig. 2.17 (Anderson and Moody, 1969). Two initial assumptions are made; the longitudinal force in the column is constant throughout its length at any given time, and the pre-curvature due to the dead-load and the lateral response of the column are both definable as half sine waves.

By considering the equilibrium of an element of the column the basic differential equation of motion can be written down

$$-\frac{\partial^2 M}{\partial x^2} + P\left(\frac{\partial^2 v_0}{\partial x^2} + \frac{\partial^2 v}{\partial x^2}\right) - q = 0, \tag{2.241}$$

where M is the total internal bending moment in the column, x is a longitudinal position coordinate, P is the total time dependent axial column load, v_0 is the unstressed lateral column displacement measured

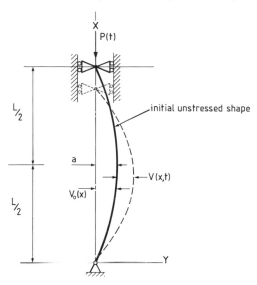

Figure 2.17 Schematic of axially loaded column with dead-load component (after Anderson and Moody, 1969). Note that there may be an additional lateral load per unit length acting which is not shown but which could be included in the form of $q(x, t)$ in equation (2.241).

from the x-axis, v is the dynamic lateral column displacement measured from the initial unstressed curve position of the column and q is included as a potential additional load per unit length which acts laterally on the column. The bending moment M may be restated by means of the moment curvature relationship

$$-M = EI \frac{\partial^2 v}{\partial x^2} \qquad (2.242)$$

(noting that EI, the product of Young's modulus E and the moment of inertia of the cross-section I about the bending axis, is the flexural rigidity). The lateral load q is likely to comprise components which are 'inherent to the structure' (lateral inertia and damping), given by

$$q(x, t) = -m\ddot{v} - c\dot{v}, \qquad (2.243)$$

where the dots have their usual meaning of time differentiation, and m denotes the mass of the column per unit length.

Substitution of equations (2.242) and (2.243) into equation (2.241) gives the following

$$\frac{\partial^2}{\partial x^2} \left(EI \frac{\partial^2 v}{\partial x^2} \right) + P \left(\frac{\partial^2 v_0}{\partial x^2} + \frac{\partial^2 v}{\partial x^2} \right) + m\ddot{v} + c\dot{v} = 0. \qquad (2.244)$$

In the original work of Anderson and Moody, equation (2.244) is non-dimensionalized by introducing the following variables, $v_0 = ru_0$; $v = ru$; $x = Ln$; $T = t\Omega$, in which r is the radius of gyration of the cross-section about the axis of bending, u_0 is a dimensionless initial unstressed lateral column displacement, u is a dimensionless dynamic lateral displacement, n is a dimensionless position coordinate, T is a dimensionless time variable and Ω is defined as the lateral frequency of vibration of a simply supported column of constant section loaded by a constant axial force P_0 such that,

$$\Omega = \frac{\pi^2}{L^2} \left[\frac{EI}{m} (1 - P_0^*) \right]_1^2, \qquad (2.245)$$

where P_0^* is the ratio of a constant axial force, P_0, to the fundamental Euler bucking load, i.e.

$$P_0^* = \frac{P_0 L^2}{\pi^2 EI}. \qquad (2.246)$$

By substituting the dimensionless variables (as defined above) and equation (2.245) as required, into equation (2.244) a dimensionless equation of motion emerges

$$u'''' + \pi^2 P^* u'' + \pi^2 P^* u_0'' + \frac{L^4 c \Omega}{EI} \dot{u} + \pi^4 \frac{\Omega^2}{\omega^2} \ddot{u} = 0. \qquad (2.247)$$

It should be noted that ω is the free lateral vibration frequency of a simply supported beam (given by equation (2.245) for the case $P_0^* = 0$). The primes denote dimensionless position derivatives. Note that

$$P^* = \frac{PL^2}{\pi^2 EI}. \qquad (2.248)$$

The initial assumptions of half sine wave curvature for the static deflection and the lateral vibration may now be used to define a time dependent displacement coordinate, therefore,

$$u_0 = (a/r) \sin(\pi n), \qquad (2.249)$$

$$u = f \sin(\pi n), \qquad (2.250)$$

a is the initial displacement of the midspan and f is a dimensionless time dependent displacement, also at the midspan. Substitution of equations (2.248)–(2.250) into equation (2.247) allows the following equation to be written

$$\ddot{f} + \frac{c}{m\Omega} \dot{f} + \frac{\omega^2}{\Omega^2} (1 - P^*)f = \frac{P^* a \omega^2}{r \Omega^2}. \qquad (2.251)$$

Substitution of equation (2.240) with subsequent rearrangement provides the final form of the governing equation such that we now have

$$\ddot{f} + 2\eta\dot{f} + \left[1 - 2\mu\sin\left(\frac{\Omega_1 T}{\Omega}\right)\right]f = \left(\frac{a}{r}\right)\left[\frac{P_0{}^*}{(1 - P_0{}^*)} + 2\mu\sin\left(\frac{\Omega_1 T}{\Omega}\right)\right].$$

(2.252)

The excitation and damping parameters are, respectively,

$$\mu = \frac{P_1{}^*}{2(1 - P_0{}^*)},$$

(2.253)

$$\eta = \frac{c}{2m\Omega}.$$

(2.254)

Equation (2.252) is an inhomogeneous Mathieu type equation with damping and a d.c. term (first right-hand side term). Neglecting the damping and both right-hand side terms reduces equation (2.252) right down to the basic Mathieu equation for which the region of principal parametric resonance will occur at $\Omega_1/\Omega = 2.0$, widening with increasing μ.

In addition to this result we can consider the whole of equation (2.252) from which the maximum lateral deflections of the column in the stable regions may be evaluated. Anderson and Moody used a first-order harmonic balance approach with the assumption of a steady-state solution of the form,

$$f = A + B\sin\left(\frac{\Omega_1 T}{\Omega}\right) + C\cos\left(\frac{\Omega_1 T}{\Omega}\right).$$

(2.255)

Substitution of this into equation (2.252), and neglecting higher harmonics, yield the following results which are in terms of the physical system parameters

$$A = 2\mu\left(\frac{a}{r}\right)\left(\frac{P_0}{P_1}\right) + \mu B,$$

(2.256)

$$B = \frac{2\mu(a/r)[1 + 2\mu(P_0/P_1)]}{[1 - 2\mu^2 - (\Omega_1/\Omega)^2] + [4\mu^2(\Omega_1/\Omega)^2]/[1 - (\Omega_1/\Omega)^2]},$$

(2.257)

$$C = \frac{-4\mu(a/r)(\Omega_1/\Omega)[1 + 2\mu(P_0/P_1)]\mu}{[1 - (\Omega_1/\Omega)^2][1 - 2\mu^2 - (\Omega_1/\Omega)^2] + 4\mu^2(\Omega_1/\Omega)^2}.$$

(2.258)

From equation (2.255) it can be seen that the maximum steady state amplitude will be

$$f_{max} = A + (B^2 + C^2)^{1/2}.$$

(2.259)

Figure 2.18 shows computer solutions for both stable steady-state response behaviour and the unstable parametric zone effect. One case of static load $P_0{}^*$ is given. It should be kept in mind that $a/r = 0.1$ here which means that the forcing term in equation (2.252) is effectively one tenth of the magnitude of the parametric excitation term. In the stable

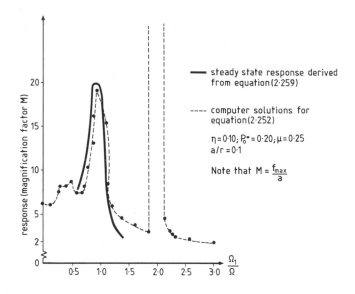

Figure 2.18 Theoretical solutions as functions of non-dimensionalized external excitation frequency (after Anderson and Moody, 1969).

region the parametric excitation is sufficiently detuned not to destabilize the system and such a case could be further investigated in a general manner in the form of the system represented by equation (2.198)–(2.200). Since there is no explicit provision for the relative detuning between the forced and the parametric excitations of the system of equation (2.252) the conclusion is that there is an exact $2:1$ ratio between parametrically driven and forced driven responses. This is clearly validated in the response plots of Fig. 2.18.

(b) A fluid–structure interaction problem

The interaction between a fluid and its container when the container is subjected to an imposed motion can be a critical aspect of structural and machine design. For example an understanding of the response of water towers to structural vibration and liquid propellant in rockets is vital if these systems are to perform well in all conditions. There is a relatively confined literature on this subject, however some useful work has been done on the parametric excitation of liquid–structure systems where the excitation function is a random process (Abramson, 1966; Dalzell, 1967; Mitchell, 1968; Ibrahim, 1972; Ibrahim and Barr, 1975). Random vibrations are largely outside the scope of this book and the reader is referred to two authoritative books on the subject for further reading

(Crandall and Mark, 1963; Ibrahim, 1985). In this example we consider the mechanics of a liquid contained within a cylindrical receptacle with an arbitrary excitation function applied to the base along the main axis of the receptacle. The problem as originally proposed (Ibrahim and Barr, 1975; Ibrahim and Soundararajan, 1983), contained a detailed analytical treatment of the system from the derived nonlinear, stochastic differential equation of motion to an averaging process through which the sloshing mode responses were outlined. We will restrict the analysis to a cursory derivation of the equation of motion in which the excitation function is arbitrary. The reader is referred to the above references (and Woodward, 1966) for further details.

Figure 2.19 shows the system in schematic form, the undisturbed free fluid surface is at height *h* and the receptacle radius is *a*.

First of all we specify a coordinate frame and for this it is appropriate to use cylindrical coordinates $Or\theta z$ fixed to the tank. Motion of the fluid is therefore relative to this. We use the standard Bernoulli equation for unsteady flow of an incompressible, inviscid fluid (which does not rotate) and apply potential flow theory to obtain, initially,

$$\Delta\Phi = 0, \tag{2.260}$$

where Δ is the Laplace operator such that

$$\Delta = \left(\frac{\partial^2}{\partial x^2} + \frac{\partial^2}{\partial y^2} + \frac{\partial^2}{\partial z^2}\right). \tag{2.261}$$

Φ is a scalar velocity potential function whose negative gradient is equal to the velocity of the fluid

$$\bar{v} = -\text{grad } \Phi = -\nabla\Phi, \tag{2.262}$$

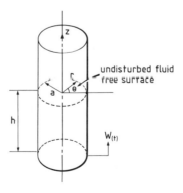

Figure 2.19 Fluid/structure interaction schematic (after Ibrahim and Barr, 1975, and Ibrahim and Soundararajan, 1983).

$$\frac{P}{\rho} + \tfrac{1}{2}\,(\nabla\Phi.\nabla\Phi) - \frac{\partial\Phi}{\partial t} + [g + \ddot{W}(t)]z = C(t). \qquad (2.263)$$

The first left-hand side term is the standard pressure term, the second term is analogous to the usual Bernoulli $v^2/2$ term after substituting equation (2.262), (note that we use the scalar product of the velocity vector $\bar{v}$ with itself). The third term is a velocity term for unsteady flow, and the final term is a composite term containing gravitational and external excitation accelerations – in essence an 'equivalent' gravitational term, z is the vertical displacement of the fluid surface from rest, and so for a free undisturbed surface we will have $z = 0$. $C(t)$ is an arbitrary time dependent function which may be subsumed into the unsteady flow velocity term.

In order to solve equation (2.263) we have to consider several boundary conditions; these are as follows:

(1) At the wall and base of the receptacle (the 'wetted' portion of the inner wall only) the normal velocity component must equal zero. Therefore,

$$\left.\frac{\partial\bar{\Phi}}{\partial r}\right|_{r=a} = \left.\frac{\partial\bar{\Phi}}{\partial z}\right|_{z=-h} = 0. \qquad (2.264)$$

(2) At the free surface (z not necessarily equal to zero) the dynamic boundary condition is that $P = 0$, therefore

$$\frac{\partial\bar{\Phi}}{\partial t} - \tfrac{1}{2}\,(\nabla\Phi.\nabla\Phi) - [g + \ddot{W}(t)]\eta = 0, \qquad (2.265)$$

(where $z = \eta$).

(3) Finally, that zero surface tension is specified by equating the vertical velocity of a fluid particle in the free surface to the vertical velocity of the free surface itself. The derivation of this boundary condition may be obtained from the literature (Dodge, Kana and Abramson, 1965). This condition is

$$\frac{\partial\eta}{\partial t} + \frac{\partial\bar{\Phi}}{\partial z} - \frac{\partial\bar{\Phi}}{\partial r}\frac{\partial\eta}{\partial r} - \left.\left(\frac{1}{r^2}\right)\frac{\partial\bar{\Phi}}{\partial\theta}\frac{\partial\eta}{\partial\theta}\right|_{z=\eta} = 0. \qquad (2.266)$$

Equation (2.260) is the continuity equation (Laplace's equation) and has a solution which satisfies the boundary condition given by equation (2.264). This solution is given in the following form

$$\bar{\Phi} = \sum_{m=0}^{\infty}\sum_{n=1}^{\infty} \alpha_{mn}J_m\cos m\theta\,\frac{\cosh[\lambda_{mn}(z+h)]}{\cosh\lambda_{mn}h} \qquad (2.267)$$

where $\alpha_{mn} = \alpha_{mn}(t)$, a time dependent coordinate obtained by satisfying boundary conditions (2.265) and (2.666), and J_m is an mth order Bessel Function of the first kind with λ_{mn} a root of $J_m' = 0$. Consideration of equations (2.260), (2.265) and (2.266) also allows the free surface elevation η to be defined,

$$\eta = \sum_{m=0}^{\infty} \sum_{n=1}^{\infty} a_{mn} J_m \cos m\theta, \tag{2.268}$$

where $a_{mn} = a_{mn}(t)$.

Equations (2.267) and (2.268) can be used to determine two further equations in terms of a_{mn} and α_{mn} by taking a Taylor series expansion of boundary conditions (2.265) and (2.266) and then a further series expansion of the nonlinear terms. The coordinate a_{mn} describes the free surface displacement amplitude

$$\alpha_{mn} - (g + \ddot{W})a_{mn} + S_{mn} = 0, \tag{2.269}$$

$$\dot{a}_{mn} + (\lambda_{mn} \tanh \lambda_{mn} h)\alpha_{mn} + Q_{mn} = 0, \tag{2.270}$$

where S_{mn} and Q_{mn} contain nonlinear terms involving α_{mn} and a_{mn} respectively. By differentiating equation (2.270) with respect to time and combining it with equation (2.269) we can obtain one equation for the 'sloshing' mode mn (for the first axisymmetric mode we have $mn = 01$),

$$\ddot{a}_{mn} + g\lambda_{mn}(\tanh \lambda_{mn} h)(1 + \ddot{W}/g)a_{mn} + F_{mn} = 0, \tag{2.271}$$

where F_{mn} contains all the nonlinear terms. $F_{mn} - F_{mn}(a_{mn}, \dot{a}_{mn}, \ddot{a}_{mn})$.

It is realistic to include a damping term here which represents the dissipation at the fluid free surface and to this end we simply add in the term $2\xi_{mn}\dot{a}_{mn}$. Defining the natural frequency of liquid sloshing as

$$\omega_{mn}^2 = g\lambda_{mn} \tanh \lambda_{mn} h \tag{2.272}$$

leads to a differential equation of motion with a time dependent coefficient

$$\ddot{a}_{mn} + 2\xi_{mn}\dot{a}_{mn} + \omega_{mn}^2(1 + \ddot{W}/g)a_{mn} + F_{mn} = 0. \tag{2.273}$$

The unspecific non-linearities contained within term F_{mn} may be neglected in a preliminary appraisal of the stability of the system.

2.4.4 Problems in applied physics

Because the problems appropriate for inclusion in this section are generally well outside the mechanical or structural engineer's experience of vibration studies it is felt that brief reviews of the phenomena will be sufficient here, and that further work on the part of the reader may follow on recourse to the cited references.

(a) Temperature driven convective instability

The stability of convection velocity in a plane, horizontal, layer of fluid subjected to a periodically varying temperature gradient is an interesting phenomenon which couples convection velocity and periodic temperature change through the well-known equations of thermal convection. This problem has been considered theoretically (Gershuni and Zhukhovitskii, 1963) and it can be shown that an equation of the damping Mathieu–Hill type will result in the case of a periodic temperature gradient. In common with the fluid–structure interaction problem of Section 2.4.3 we start off with a plane horizontal fluid layer bounded by the planes $z = \pm h$, the z-axis being vertically orientated. At thermal equilibrium the temperature $T_0 = T_0(z, t)$ satisfies the unsteady heat conduction equation

$$\frac{\partial T_0}{\partial t} = K \frac{\partial^2 T_0}{\partial z^2} \qquad (2.274)$$

and the velocity $v = 0$; K is the coefficient of heat conduction. The imposed heating condition is that the equilibrium temperature gradient in the fluid varies with frequency ω_0 about some mean value but that ω_0 is within a regime of fairly low frequencies defined by

$$\omega_0 \ll \frac{K}{h^2}. \qquad (2.275)$$

So the equilibrium temperature gradient does not depend on z but on the following general function

$$\frac{\partial T_0}{\partial z} = -A_0 + a_0 \psi(t), \qquad (2.276)$$

where A_0 is the mean temperature gradient, a_0 is the temperature variation amplitude (or modulation depth) and $\psi(t)$ is a periodic time varying function at frequency ω_0. This could be $\cos \omega_0 t$ for example.

The convection equation can now be stated. Note that a full derivation of these is readily available (Jaluria, 1980)

$$\frac{\partial v}{\partial t} = -\frac{1}{\rho} \nabla P + v \Delta v = g \beta T, \qquad (2.277)$$

$$\frac{\partial T}{\partial t} + v_z \frac{\partial T_0}{\partial z} = K \Delta T, \qquad (2.278)$$

for which we define ρ as the fluid density, P is the pressure, v is the kinematic viscosity, $\beta = -(1/\rho)(\partial \rho / \mathrm{d}T)$, v_z is the axial convection velocity component, and T is the absolute temperature, ∇ is the grad function and Δ is the Laplace operator of equation (2.261). It is clear that subsequent coupling between temperature and fluid velocity is initiated in equations (2.277) and (2.278). By taking small perturbations for v_z and T and by introducing new velocity and temperature functions, $v(t)$ and $\tau(t)$, which are solely dependent upon time,

$$\dot{v}_z = v(t) \cos\left(\frac{\pi}{2h}\right) z, \tag{2.279}$$

$$T = \tau(t) \cos\left(\frac{\pi}{2h}\right) z, \tag{2.280}$$

two further equations may be developed from equations (2.277) and (2.278)

$$v + vK_1^2 v = \frac{g\beta c^2}{K_1^2} \tau, \tag{2.281}$$

$$\tau + KK_1^2 \tau = -T_0' v, \tag{2.282}$$

$(K_1^2 = c^2 + \pi^2/4h^2$; where c is an arbitrary system constant).

The dots and primes denote differentiation with respect to time t and axial position z respectively. By substituting for T_0' from equation (2.276) and eliminating $\tau(t)$ we obtain,

$$\ddot{v} + K_1^2 K \left(1 + \frac{v}{K}\right) \dot{v} + \left[\frac{g\beta c^2}{K_1^2} (a_0 \psi(t) - A_0) + vKK_1^4\right] v = 0. \tag{2.283}$$

Although this is rather unwieldy in its present form it is evidently a Mathieu–Hill type of equation with damping (second term). Rewriting this

$$\ddot{v} + C\dot{v} + [B + D\psi(t)]v = 0, \tag{2.284}$$

where

$$B = \left(vKK_1^4 - \frac{g\beta c^2 A_0}{K_1^2}\right), \tag{2.285}$$

$$C = KK_1\left(1 + \frac{v}{K}\right), \tag{2.286}$$

$$D = \frac{g\beta c^2 a_0}{K_1^2}. \tag{2.287}$$

Thus, regions of instability may be generated dependent upon the numerical data used.

(b) A brief review of applications in electronics

Parametric phenomena have been observed in physical systems which are not straightforwardly mechanical or structural in nature, for example periodic variations in capacitance and inductance are effects exploited in the design of electronic parametric amplifiers for high signal-to-noise ratio communication systems. Amplifiers of this sort fall into two categories, the up-converters and the negative resistance parametric amplifiers (or degenerate amplifiers). Although diodes which can exhibit a time variant capacitance are generally utilized in these circuits the governing equations are usually concerned with power-flow into and out of an ideal nonlinear reactance (capacitative or inductive reactance, x_c or x_i) rather than stability considerations. The fact that the most practical designs utilize ideal, stable, nonlinear elements is the reason behind this. Rather than the Mathieu–Hill type of equation, therefore, it is more usual to encounter the Manley–Rowe power-flow equations (Blackwell and Kotzebue, 1961). The up-converter has two important characteristics, these being that the output frequency is equal to the sum of the input frequency and the pump frequency (the pump provides a control over power flow within the circuit), and so that there is no power flow at frequencies other than these (it is therefore a resonant system). There is a demonstrable two-to-one ratio between pump output frequencies. The degenerate amplifier is a further refinement of the up-converter and admits the possibility of signal regeneration by means of the negative resistance present within the system. The negative resistance is usually thought of as a power generator within the amplifier.

Ibrahim (1978) points out that parametric excitations in mechanical systems coupled to magnetic or electric systems are possible and cites several references on these themes, in particular, parametric magneto-elastic resonance of a perfect elastic conductor in a magnetic field (Kaliski, 1968) and the interaction between a time dependent magnetic field and an elastic beam-plate which could go into an unstable vibration mode under certain magnetic field fluctuation frequencies (Moon and Pao, 1969).

2.4.5 Further reading

In addition to the references given in the preceding section the reader is recommended to investigate the following, many of which are classic texts and which offer detailed and comprehensive research bibliographies and reference listings. Chronologically these are (Andronov and Chaikin, 1949; Lasalle and Lefshetz, 1961; Bogoliubov and Mitropolsky, 1961; Struble, 1962; Minorsky, 1962; Hayashi, 1964; Arscott, 1964; Bolotin, 1964; Nayfeh, 1973; Evan–Iwanowski, 1976; Ibrahim and Barr, 1978; Nayfeh and Mook, 1979; Ibrahim, 1985; Schmidt and Tondl, 1986).

3

Nonlinear vibrations in forced and parametrically excited systems

INTRODUCTION

Non-linearities can appear in most real systems and their presence in one form or another is generally the rule rather than the exception. For the purposes of subsequent explanations it might be useful to attempt to group them into categories. First, we could consider the category containing non-linearities that arise because of reasonably careful mathematical modelling, in which terms are retained in their derived form. Thus we could accommodate large deflection problems this way, where the non-linearities are not based on a known physical phenomenon (such as damping) but are essentially geometrical in origin. It will be shown that non-linearities of this type, although generally considered to be relatively weak as compared with linear terms or material and configuration non-linearities, can generate interesting modifications to the responses of parametric and forced oscillators. It should be pointed out that it is common practice to eradicate such terms by linearizing the problem, and then to consider only the potential linear response behaviour. Such a restricted outlook does not always provide a satisfactory and complete answer to the problem.

Next, there is a group of non-linearities that are present because of very definite and identifiable physical characteristics, such as nonlinear damping which may be generated by material composition, and/or a very specific structural configuration. In these cases the effects upon the response are usually important enough to mean that linearization of the equations would preclude any serious understanding of the system.

It is clear, therefore, that the specific effects of non-linearities depend considerably on their individual forms and so it is important that some examples of the above are investigated in order to get an overall feel for the subject.

3.1 LARGE DEFLECTION NON-LINEARITIES

3.1.1 Forced nonlinear oscillator with a cubic non-linearity

In this section we propose to examine the influence of a cubic term on the response behaviour of a forced, single degree of freedom, oscillator. The equation of interest is as follows

$$\ddot{x} + 2\xi\omega_0\dot{x} + \omega_0^2 x \pm hx^3 = f_0\cos\Omega t. \qquad (3.1)$$

The sign of the cubic term governs the hardening or softening nature of the elastic element within the system. Therefore if we have $+hx^3$ the stiffness hardens up with large x, and conversely the spring tends to soften (with large x) for the case where we have $-hx^3$. Therefore in terms of the response the overall effect of $\pm hx^3$ is to bend the resonance curve over to one side. In the hardening spring example the curve bends over to the right (Fig. 3.4). A clear treatment of the forced problem (based on equation (3.1)) has been given by Roberts (1984) and it is this work that is to be briefly reviewed here.

The physical basis for this work is a clamped–clamped beam structure which is subjected to seismic (i.e. foundation) excitation; refer to Fig. 3.1. It can be shown that the instantaneous strain energy for the beam

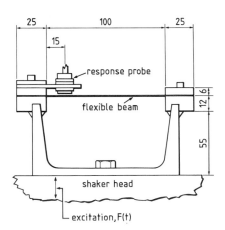

Figure 3.1 Clamped–clamped beam model for investigating a forced non-linear oscillator with a cubic nonlinearity (after Roberts, 1984).

may be given by the following

$$U = \frac{EI}{2} \int_0^l w''^2 \, dx + \frac{EAl}{8} \left[\int_0^l w'^2 \, dx \right]^2 \tag{3.2}$$

in which the first term represents the flexural strain energy, and the second term arises because of elastic axis stretching (Roberts, 1984). We assume a single mode of vibration, and a response $w(x, t)$, which is a function of the spatial and temporal coordinates $g(x)$ and $q(t)$. Therefore,

$$x(x, t) = g(x)q(t). \tag{3.3}$$

Non-dimensionalizing equation (3.3) by dividing through by some arbitrary reference displacement a leads to

$$X = g(l/2)q/a, \tag{3.4}$$

where X is the midspan response of the beam. We can therefore write

$$\ddot{X} + 2\xi\omega_0\dot{X} + \omega_0^2(X + \beta X^3) = P_0 \cos \Omega t \tag{3.5}$$

taking the hardening $(+\beta X^3)$ case only.

Equations (3.1) and (3.5) are forms of the very well-known Duffing equation, an equation which is of great interest because of its apparent simplicity of structure, yet it is capable of a rich (and in some circumstances complex) range of response characteristics. In addition to displaying response harmonics, Duffing type systems frequently display nonlinear jump phenomena and hysteresis effects (to be outlined later in this section); they are also a source of possible chaotic responses, more of which in Chapter 5.

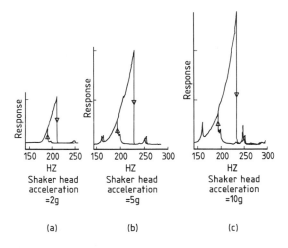

Figure 3.2 Experimental nonlinear (cubic) response for three excitation levels (after Roberts, 1984).

Experimental work by Roberts, Bux and Cartmell (Roberts, 1984) shows that the nonlinear aspect to the problem is very clearly in evidence in the responses with several cases presented in Figs 3.2 and 3.3. Figure 3.2(a)–(c) shows how the configuration of Fig. 3.1 responds to three progressively higher excitational levels. The hysteresis effect is clearly visible on all three, with the downward jump at the largest solution getting more significant as the excitation acceleration value is increased (Fig. 3.2(c)). Characteristic upward jumps are also shown, and in practice these would be realizable for a decreasing frequency sweep. The downward jumps (which are larger than the upward jumps – typical of a hardening spring system) are observable in practice by taking an increasing frequency sweep over the response region. Therefore frequencies may be identified at which two stable solutions exist. We can force the system to transfer from one solution to another simply by physically perturbing it in some way. The downward jump is probably easier to initiate, and all that would be required is the application of some sort of restriction to motion, preferably at or near the antinode; refer to Fig. 3.1. This would occur at the middle of the span of the beam in Fig. 3.1.

In Fig. 3.3 the excitation frequency is fixed at the upper jump frequency and the excitation acceleration is varied. Two responses are apparent over a particular range of accelerations with very large jumps occurring at each extreme end of the region (points *A* and *B* on Fig. 3.3).

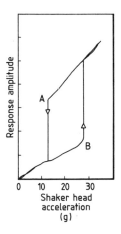

Figure 3.3 Experimental responses as a function of excitation level showing downward (i.e. from A) and upward (i.e. from B) jumps between upper and lower solutions. Note that the excitation frequency is fixed throughout (after Roberts, 1984).

Analysis of equation (3.5) will allow us to deduce an appropriate solution such that we can initially propose that

$$X(t) = A \cos(\Omega t + \psi) \tag{3.6}$$

for which it can be shown that the response amplitude A is given by the roots of

$$A\omega_0^2(1 + \tfrac{3}{4}\beta A^2) = A\Omega^2 \pm P_0 \sqrt{1 - (2\xi\omega_0\Omega(A/P_0))^2} \tag{3.7}$$

(Roberts, 1984).

Numerical solution of equation (3.7) over a range of Ω values will give one or three real roots, dependent on Ω, for the system. A selection of response curves which demonstrate this phenomenon are presented in Fig. 3.4. In the case of the main curve (bold line in Fig. 3.4) hysteresis is apparent in the range $1.223 < (\Omega/\omega_0) < 1.510$ (the precise values being closely related to specific system parameters).

This means that we can repetitively travel round the loop ABCD by sweeping up and down in frequency over this range.

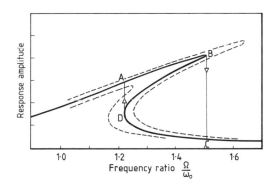

Figure 3.4 Theoretical nonlinear responses for the clamped–clamped beam under seismic excitation. The broken lines denote response paths for cases of lower and higher excitation magnitude (after Roberts, 1984).

3.1.2 Large deflection non-linearity in a simple parametrically excited system

Figure 3.5 illustrates a seemingly simple problem comprising a bobbed pendulum excited into motion by a moving support. This is a classical problem (albeit with limited practical significance) but is included here to show how important nonlinear terms can be under certain circumstances. The system is governed by the damped form of the Mathieu–

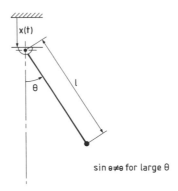

Figure 3.5 Pendulum with lumped end mass undergoing support excitation.

Hill equation, where the pendulum swing angle coordinate is θ,

$$\ddot{\theta} + 2\xi\omega_0\dot{\theta} + (\omega_0{}^2 + \ddot{x}/l)\sin\theta = 0, \tag{3.8}$$

where ξ is the damping coefficient, x is the support displacement and l is the length of the pendulum.

We could immediately reduce this to a simple linear problem by stating $\sin\theta \simeq \theta$, and therefore one for which the system is restricted to swing angles of $<15°$ to $20°$. However if we prefer to admit $\theta > 20°$ or so then $\sin\theta \neq \theta$ and we find that the problem is a little more complicated.

The Maclaurin series for $\sin\theta$ may be used here

$$\sin\theta = \theta - \frac{\theta^3}{3!} + \frac{\theta^5}{5!} - \ldots + \ldots. \tag{3.9}$$

Assuming that quintic and higher order terms will entail a disproportionate amount of algebra (compared with their overall effects) we truncate the series after $0(\theta^3)$ and substitute for $\sin\theta$ in equation (3.8). Therefore we get

$$\ddot{\theta} + 2\xi\omega_0\dot{\theta} + \left[\omega_0{}^2 + \frac{\ddot{x}}{l}\right]\left(\theta - \frac{\theta^3}{3!}\right) = 0. \tag{3.10}$$

Expanded this gives the following,

$$\ddot{\theta} + 2\xi\omega_0\dot{\theta} + \omega_0{}^2\theta - \frac{\omega_0{}^2}{6}\theta^3 + \frac{\ddot{x}\theta}{l} - \frac{\ddot{x}\theta^3}{6l} = 0. \tag{3.11}$$

We can now proceed to investigate the stability of this equation, and to do this we once again use the method of multiple scales to expand equation (3.11) into perturbation equations (up to and including $0(\varepsilon^1)$)

will be sufficient), but before this procedure is carried out the equation
needs to be ordered,

$$\ddot{\theta} + 2\varepsilon\zeta\omega_0\dot{\theta} + \omega_0{}^2\theta - \varepsilon h\theta^3 - \varepsilon f\cos\Omega t.\theta + \frac{\varepsilon f}{6}\cos\Omega t.\theta^3 = 0 \quad (3.12)$$

for which we have assumed the following forms,

$$x = X_0\cos\Omega t; \quad \ddot{x} = -X_0\Omega^2\cos\Omega t; \quad \frac{\omega_0{}^2}{6} = \varepsilon h; \quad \frac{X_0\Omega^2}{l} = \varepsilon f; \quad \xi = \varepsilon\zeta.$$

The zeroth and first-order perturbation equations are, respectively,

$$\varepsilon^0: \quad D_0{}^2\theta_0 + \omega_0{}^2\theta_0 = 0, \quad (3.13)$$

$$\varepsilon^1: \quad D_0{}^2\theta_1 + \omega_0{}^2\theta_1 = -2D_0D_1\theta - 2\zeta\omega_0D_0\theta_0$$

$$+ h\theta_0{}^3 + f\cos\Omega t.\theta_0 - \frac{f}{6}\cos\Omega t.\theta_0{}^3. \quad (3.14)$$

The solution to equation (3.13) is

$$\theta_0 = A(T_1)\exp(i\omega_0 T_0) + \bar{A}(T_1)\exp(-i\omega_0 T_0). \quad (3.15)$$

Also we can write

$$\cos\Omega t = \tfrac{1}{2}(\exp(i\Omega T_0) + \exp(-i\Omega T_0)), \quad (3.16)$$

(where $t \simeq T_0$).

Substituting equations (3.15), (3.16) in to (3.14) leads to

$$D_0{}^2\theta_1 + \omega_0{}^2\theta_1 = -2D_0D_1[A\exp(i\omega_0 T_0) + \bar{A}\exp(-i\omega_0 T_0)]$$

$$-2\zeta\omega_0 D_0[A\exp(i\omega_0 T_0) + \bar{A}\exp(-i\omega_0 T_0)]$$

$$+ h[A\exp(i\omega_0 T_0) + \bar{A}\exp(-i\omega_0 T_0)]^3$$

$$+ \frac{f}{2}[\exp(i\Omega t) + \exp(-i\Omega t)][A\exp(i\omega_0 T_0)$$

$$+ \bar{A}\exp(-i\omega_0 T_0)]$$

$$- \frac{f}{12}[\exp(i\Omega t) + \exp(-i\Omega t)][A\exp(i\omega_0 T_0)$$

$$+ \bar{A}\exp(-i\omega_0 T_0)]^3. \quad (3.17)$$

On expansion we find that this becomes

$$D_0{}^2\theta_1 + \omega_0{}^2\theta_1 = \exp i\omega_0 T_0[-i2D_1A\omega_0 + i2D_1\bar{A}\exp(-i2\omega_0 T_0)$$

$$- i2\zeta\omega_0{}^2 A + i2\zeta\omega_0{}^2\bar{A}\exp(-i2\omega_0 T_0)$$

$$+ hA^3\exp(i2\omega_0 T_0)$$

$$+ h\bar{A}^3\exp(-i4\omega_0 T_0) + 3hA\bar{A}^2\exp(-i2\omega_0 T_0)$$

$$+ 3h\bar{A}A^2$$

$$+ \frac{f}{2}A \exp(i\Omega T_0) + \frac{f}{2}\bar{A} \exp i(\Omega - 2\omega_0)T_0$$

$$+ \left(\frac{f}{2}\right)A \exp(-i\Omega T_0)$$

$$+ \frac{f}{2}\bar{A} \exp i(-\Omega - 2\omega_0)T_0$$

$$- \frac{f}{12}A^3 \exp i(\Omega + 2\omega_0)T_0$$

$$- \frac{f}{12}\bar{A}^3 \exp i(\Omega - 4\omega_0)T_0$$

$$- \frac{f}{4}A\bar{A}^2 \exp i(\Omega - 2\omega_0)T_0$$

$$- \frac{f}{4}\bar{A}A^2 \exp(i\Omega T_0) - \frac{f}{12}A^3 \exp i(2\omega_0 - \Omega)T_0$$

$$- \frac{f}{12}\bar{A}^3 \exp i(-\Omega - 4\omega_0)T_0$$

$$- \frac{f}{4}A\bar{A}^2 \exp i(-\Omega - 2\omega_0)T_0$$

$$- \frac{f}{4}A^2\bar{A} \exp(-i\Omega T_0)]. \tag{3.18}$$

Removal of secular generating terms leads to

$$-i2D_1A\omega_0 - i2\zeta\omega_0^2A + 3h\bar{A}A^2 + \frac{f}{2}\bar{A} \exp i(\Omega - 2\omega_0)T_0$$

$$- \frac{f}{4}A\bar{A}^2 \exp i(\Omega - 2\omega_0)T_0 - \frac{f}{12}A^3 \exp i(2\omega_0 - \Omega)T_0 = 0. \tag{3.19}$$

We let

$$A = \frac{a}{2}\exp i\alpha; \quad \bar{A} = \frac{a}{2}\exp(-i\alpha); \quad a = a(T_1); \quad \alpha = \alpha(T_1)$$

therefore

$$D_1A = \frac{a'}{2}\exp i\alpha + \frac{a}{2}i\alpha' \exp i\alpha. \tag{3.20}$$

Substitutions for A, $\bar{A}$ and D_1A in equation (3.19) lead to

$$-i\omega_0a' \exp i\alpha + \omega_0aa' \exp i\alpha - i\zeta\omega_0^2a \exp i\alpha + \tfrac{3}{8}ha^3 \exp(i\alpha)$$

$$+ fa \exp i[(\Omega - 2\omega_0)T_0 - \alpha]$$

$$- \frac{f}{32}a^3 \exp i[(\Omega - 2\omega_0)T_0 - \alpha]$$

$$-\frac{fa^3}{96}\exp \mathrm{i}[(2\omega_0 - \Omega)T_0 + 3\alpha] = 0. \qquad (3.21)$$

Substituting for the principal parametric resonance case ($\Omega = 2\omega_0 + \varepsilon\sigma$) gives

$$-\mathrm{i}\omega_0 a' + \omega_0 a\alpha' - \mathrm{i}\zeta\omega_0^2 a + (\tfrac{3}{8})ha^3 + fa\exp \mathrm{i}(\varepsilon\sigma T_0 - 2\alpha)$$

$$-\frac{f}{32}a^3 \exp \mathrm{i}(\varepsilon\sigma T_0 - 2\alpha) - \frac{fa^3}{96}\exp \mathrm{i}(2\alpha - \varepsilon\sigma T_0) = 0. \qquad (3.22)$$

Separation out into real and imaginary parts gives

$$\text{Re:}\quad \omega_0 a\alpha' + \frac{3ha^3}{8} + fa\cos(\varepsilon\sigma T_0 - 2\alpha) - \frac{f}{32}a^3\cos(\varepsilon\sigma T_0 - 2\alpha)$$

$$-\frac{fa^3}{96}\cos(2\alpha - \varepsilon\sigma T_0) = 0, \qquad (3.23)$$

$$\text{Im:}\quad -\omega_0 a' - \zeta\omega_0^2 a + fa\sin(\varepsilon\sigma T_0 - 2\alpha) - \frac{fa^3}{32}\sin(\varepsilon\sigma T_0 - 2\alpha)$$

$$-\frac{fa^3}{96}\sin(2\alpha - \varepsilon\sigma T_0) = 0, \qquad (3.24)$$

which, on rearrangement, gives

$$\omega_0 a\alpha' + \frac{3ha^3}{8} + fa\left(1 - \frac{a^2}{32} - \frac{a^2}{96}\right)\cos\phi = 0, \qquad (3.25)$$

$$-\omega_0 a' - \zeta\omega_0^2 a + fa\left(1 - \frac{a^2}{32} + \frac{a^2}{96}\right)\sin\phi = 0, \qquad (3.26)$$

where

$$\phi = \varepsilon\sigma T_0 - 2\alpha \qquad (= \sigma T_1 - 2\alpha). \qquad (3.27)$$

The prime denotes differentiation with respect to slow time scale T_1 therefore

$$a' \simeq \phi' \simeq 0, \quad \text{giving}\quad \sigma = 2\alpha'. \qquad (3.28)$$

Substituting equation (3.28) into (3.25) eliminates α'

$$\omega_0\frac{\sigma}{2} + \frac{3ha^2}{8} + f\left(1 - \frac{a^2}{32} - \frac{a^2}{96}\right)\cos\phi = 0. \qquad (3.29)$$

Also

$$\zeta\omega_0^2 + f\left(1 - \frac{a^2}{32} + \frac{a^2}{96}\right)\sin\phi = 0. \qquad (3.30)$$

Squaring and adding these two equations yields

$$\frac{(\omega_0(\sigma/2) + 3ha^2/8)^2}{(1 - a^2/32 - a^2/96)^2} + \frac{\zeta^2\omega_0^4}{(1 - a^2/32 + a^2/96)^2} = f^2. \qquad (3.31)$$

Neglecting terms of order greater than four in a, and multiplying through by ε^2 gives the following quadratic approximation for a^2,

$$\left[\frac{187}{2304}(\varepsilon f)^4 + \frac{\omega_0}{192}\varepsilon\sigma(\varepsilon f)^2\left(\frac{\omega_0}{48}\varepsilon\sigma - 3\varepsilon h\right) + \frac{\omega_0{}^4}{576}\xi^2(\varepsilon f)^2 + \tfrac{9}{64}(\varepsilon h)^2(\varepsilon f)^2\right]a^4$$

$$+ \left[\frac{(\varepsilon f)^2}{8}\left(\frac{(\varepsilon f)^2}{3} + 3\varepsilon h\varepsilon\sigma\omega_0\right) - \frac{\omega_0{}^2}{12}(\varepsilon f)^2\left(\frac{(\varepsilon\sigma)^2}{8} + \omega_0{}^2\xi^2\right)\right]a^2$$

$$+ (\varepsilon f)^2\left[\omega_0{}^2\left(\frac{(\varepsilon\sigma)^2}{4} + \xi^2\omega_0{}^2\right) - (\varepsilon f)^2\right] = 0. \tag{3.32}$$

We can now resubstitute the original definitions for εh, εf, $\varepsilon\zeta$ back into equation (3.32).

$$\left[\frac{187}{2304}\left(\frac{X_0\Omega^2}{l}\right)^4 + \frac{\omega_0{}^2\varepsilon\sigma}{384}\left(\frac{X_0\Omega^2}{l}\right)^2\left(\frac{\varepsilon\sigma}{24} - \omega_0\right)\right.$$

$$\left. + \frac{\omega_0{}^4\xi^2}{587}\left(\frac{X_0\Omega^2}{l}\right)^2 + \frac{9\omega_0{}^4}{2304}\left(\frac{X_0\Omega^2}{l}\right)^2\right]a^4$$

$$+ \left[\tfrac{1}{8}\left(\frac{X_0\Omega^2}{l}\right)^2 (\tfrac{1}{3})\left(\frac{X_0\Omega^2}{l}\right)^2 + (\tfrac{1}{2})\omega_0{}^3\varepsilon\sigma\right)$$

$$- (\tfrac{1}{12})\omega_0{}^2\left(\frac{X_0\Omega^2}{l}\right)^2((\tfrac{1}{8})(\varepsilon\sigma)^2 + \omega_0{}^2\xi^2)\right]a^2$$

$$+ \left(\frac{X_0\Omega^2}{l}\right)^2\left[\omega_0{}^2((\tfrac{1}{4})(\varepsilon\sigma)^2 + \xi^2\omega_0{}^2) - \left(\frac{X_0\Omega^2}{l}\right)^2\right] = 0. \tag{3.33}$$

Equation (3.33) describes the relationship between excitation amplitude, detuning, and system response.

Figure 3.6 illustrates the characteristic responses of the system described by equation (3.33) and it is clear how the inclusion of $\sin\theta$ in the governing equation (rather than θ) demonstrably affects the qualitative nature of the responses. Four cases are given; these cover two different excitation accelerations and two different damping ratios. In general it can be seen that the curves bend over to the left of the $\varepsilon\sigma-a$ region; a characteristic commonly described as that of a softening spring where the cubic term is negative, as it is in equation (3.12). If we take the case for which $X_0\Omega^2 = 10$; $\xi = 0.001$, as an example it can be seen that for $\varepsilon\sigma = -4$ there are two predicted solutions which are bounded. The upper solution, A (upper branch of the curve), is stable whilst the lower one, B, is not. Had the linearized problem of $\sin\theta = \theta$ been considered all solutions within the region would be unbounded (non-steady-state and therefore unstable). Therefore the act of linearization

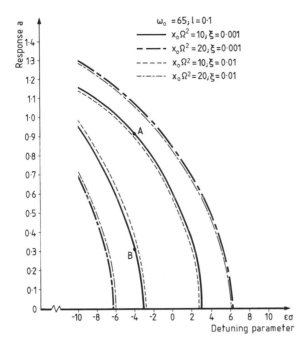

Figure 3.6 Softening spring characteristic for the parametrically excited pendulum; various cases.

could completely distort the statement of the problem since it precludes the case of steady-state solutions. Clearly it is a technique that should be handled carefully.

3.1.3 Geometrical non-linearities in parametrically excited systems

(a) Kinematics of a parametrically excited cantilever structure

The base excited cantilever beam of Fig. 2.5(d) may be reconsidered, and the linearized equations (2.201) to (2.203) restated in a fuller form with many geometrical non-linearities included. The nonlinear terms all have their source in the system kinetic energy, given below for three modal coordinates, u_{01}, u_{02}, ϕ_0,

$$T = \tfrac{1}{2}m_0[\dot{u}_{01}{}^2 + \dot{u}_{02}{}^2 + \dot{w}_0{}^2 + (\dot{v}_0 + \dot{W})^2] + \tfrac{1}{2}I_0\dot{\phi}_0{}^2. \quad (3.34)$$

For combined bending and torsion there has to be a small but highly significant in-plane displacement v_0 (in-plane refers to to the excitational plane OY), and it is this which generates many of the possible non-linearities. The arbitrarily chosen lateral displacement coordinate u

(Figure 2.5(d)) may be related to the modal coordinates of lateral displacement u_{01}, and u_{02} (these are deemed to define lateral translations at the centre of the lumped mass m_0), therefore,

$$u(z, t) = f_1(z)u_{01}(t) + f_2(z)u_{02}(t), \qquad (3.35)$$

where $f_1(z)$ and $f_2(z)$ are the linear fundamental and second bending mode shapes. In order to express v_0 in terms of the modal coordinates it is necessary to examine the kinematics of the system in a certain amount of detail. The kinematical foundation, on which the following work is built, is the Euler–Kirchoff–Love representation for rods (Love, 1944) applied to a cantilever beam.

We can begin by considering an element of a slender rectangular section beam as shown in Fig. 3.7, and can assign a three-dimensional frame of reference to it in the undeformed state which we call $OXYZ$. For all intents and purposes this frame is completely fixed in space and so we can, in a similar manner, propose another frame, $Oxyz$ which defines the spatial position of the element once it has undergone some arbitrary displacement in bending and torsion. Therefore we have a basis for defining the position in space of the element as it moves from A to B. Each frame also possesses a triad of unit vectors of which more later.

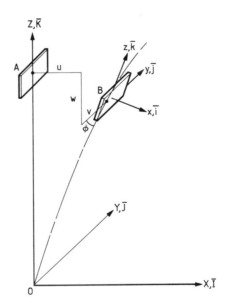

Figure 3.7 Coordinate reference frames for a beam element.

The change in position from XYZ to xyz can be examined in terms of component displacements u, v, w and a twist angle ϕ. The lateral displacement of the element is represented by u in the Oxz plane, w is the vertical 'drop' associated with bending in this plane and v is a small displacement in the Oyz plane brought about by combined bending (u) and torsion (ϕ), where ϕ is the angle of twist of the element around the z-axis.

Having established these representations we can go on to describe combined bending and torsion of the element through a series of successive rotations of an orthogonal frame which begins at $OXYZ$ and ends at $Oxyz$. The passage of the frame from $OXYZ$ to $Oxyz$ is made through several discrete stages (and through several discrete angles). These angles are generally known as Euler angles. In Fig. 3.8 the three rotations of the reference frame are shown, firstly about axis OY in which case the frame $OXYZ$ rotates through α to $OX_1Y_1Z_1$. Subsequently the frame rotates through β, about OX_1, to get to $OX_2Y_2Z_2$, and finally through ϕ, about OZ_2, to arrive at $Oxyz$. We have chosen three rotations to correspond with the three dimensions of the frame. Figure 3.8 is useful in that it gives a physical foundation for what follows when we attempt to describe the positional change experienced by the element in terms of mathematical equations.

From Fig. 3.8 we can say that the relationship between X and X_1 is $\cos \alpha$, i.e. $X_1 = X \cos \alpha$, and since the repositioning of Z to Z_1 is in the same plane (but involving an additional $90°$) we have $X_1 = Z \cos(90 + \alpha)$. There is no planar relationship between X_1 and Y, X and Y_1, Z and Y_1, however, since the rotation through α is about OY then $OY = OY_1$, thus $Y = Y_1$. Examination of Fig. 3.8 should now help to illustrate how we get to $Z_1 = X \cos(90 - \alpha)$ and $Z_1 = Z \cos \alpha$. This can be summarized in the following form,

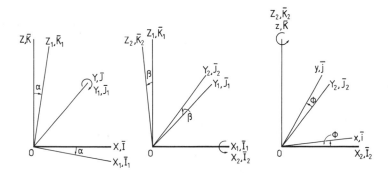

Figure 3.8 Coordinate frame rotations through Euler angles α, β and ϕ respectively.

	X	Y	Z
X_1	$\cos \alpha$	0	$\cos(90 + \alpha)$
Y_1	0	1	0
Z_1	$\cos(90 - \alpha)$	0	$\cos \alpha$

The entries in the body of the table are usually called direction cosines and can be put into a direction cosine matrix

$$A = \begin{bmatrix} \cos \alpha & 0 & -\sin \alpha \\ 0 & 1 & 0 \\ \sin \alpha & 0 & \cos \alpha \end{bmatrix}. \tag{3.36}$$

A similar analysis of the rotations about β and ϕ leads to

$$B = \begin{bmatrix} 1 & 0 & 0 \\ 0 & \cos \beta & -\sin \beta \\ 0 & \sin \beta & \cos \beta \end{bmatrix} \tag{3.37}$$

and

$$C = \begin{bmatrix} \cos \phi & \sin \phi & 0 \\ -\sin \phi & \cos \phi & 0 \\ 0 & 0 & 1 \end{bmatrix}. \tag{3.38}$$

Unit vectors may be associated with each reference frame axis, so that $\bar{I}, \bar{J}, \bar{K}$ are allied with X, Y, Z and $\bar{I}_1, \bar{J}_1, \bar{K}_1$ with X_1, Y_1, Z_1, and $\bar{I}_2, \bar{J}_2, \bar{K}_2$ with X_2, Y_2, Z_2. Finally we could have $\bar{i}, \bar{j}, \bar{k}$ assigned to x, y, z. The introduction of unit vectors allows a formal representation of whole frame motion to be developed, and so since the three successive frame rotations are about OY, OX_1, and OZ_2 respectively (with unit vectors J, I_1 and K_2 respectively) we can write down the vector sum of the rates of rotation of the frame as it moves from $OXYZ$ to $Oxyz$

$$\bar{\omega} = \frac{d\alpha}{dt} \bar{J} - \frac{d\beta}{dt} \bar{I}_1 + \frac{d\phi}{dt} \bar{K}_2, \tag{3.39}$$

where ω is the rate of rotation vector for the whole frame as it progresses from $OXYZ$ to $Oxyz$. The second term is negative because the rotation is negative in the sense of the right-hand screw rule (for a right-hand orthogonal vector set such as we have here).

In order to do any more with equation (3.39) it is necessary to find a way of rewriting it so that it is in terms of those unit vectors that relate to one reference frame, preferably $Oxyz$. We can accomplish this by manipulating the direction cosine matrices so that, for example, we can relate $\bar{J}$ to $\bar{i}, \bar{j}, \bar{k}$. To achieve this step we must multiply the three direction cosine matrices together (as this transformation is commensurate with $OXYZ$ moving to $Oxyz$, i.e. $\bar{I}, \bar{J}, \bar{K}$ to $\bar{i}, \bar{j}, \bar{k}$). Therefore we obtain the following,

	$X, \bar{I}$	$Y, \bar{J}$	$Z, \bar{K}$
$x, \bar{i}$	$(\cos\phi\cos\alpha$ $- \sin\phi\sin\beta\sin\alpha)$	$\sin\phi\cos\beta$	$(-\cos\phi\sin\alpha$ $- \sin\phi\sin\beta\cos\alpha)$
$y, \bar{j}$	$(-\sin\phi\cos\alpha$ $- \cos\phi\sin\beta\sin\alpha)$	$\cos\phi\cos\beta$	$(\sin\phi\sin\alpha$ $- \cos\phi\sin\beta\cos\alpha)$
$z, \bar{k}$	$(\cos\beta\sin\alpha)$	$\sin\beta$	$\cos\beta\cos\alpha$

It is clear from the table that

$$\bar{J} = \bar{i}\sin\phi\cos\beta + \bar{j}\cos\phi\cos\beta + \bar{k}\sin\beta. \tag{3.40}$$

Returning to equation (3.39) we now need $\bar{I}_1$ in terms of $\bar{i}, \bar{j}, \bar{k}$, which is slightly simpler than the previous stage because these vectors are associated with $OX_1Y_1Z_1$ moving through to $Oxyz$ (or through β and ϕ). The product $B.C$ (equations (3.37) and (3.38)) satisfies this requirement giving

	$X_1, \bar{I}_1$	$Y_1, \bar{J}_1$	$Z_1, \bar{K}_1$
$x, \bar{i}$	$\cos\phi$	$\sin\phi\cos\beta$	$-\sin\phi\sin\beta$
$y, \bar{j}$	$-\sin\phi$	$\cos\phi\cos\beta$	$-\sin\beta\cos\phi$
$z, \bar{k}$	0	$\sin\beta$	$\cos\beta$

Therefore,

$$\bar{I}_1 = \bar{i}\cos\phi - \bar{j}\sin\phi. \tag{3.41}$$

Finally K_2 in terms of i, j, k, may be found by considering the frame rotation through ϕ on its own (refer to equation (3.38)

	$X_2\bar{I}_2$	$Y_2\bar{J}_2$	$Z_2\bar{K}_2$
$x, \bar{i}$	$\cos\phi$	$\sin\phi$	0
$y, \bar{j}$	$-\sin\phi$	$\cos\phi$	0
$z, \bar{k}$	0	0	1

And so we have

$$\bar{K}_2 = \bar{i}(0) + \bar{j}(0) + \bar{k}(1) = \bar{k}. \tag{3.42}$$

Substituting equations (3.40), (3.41), (3.42) into equation (3.39) gives

$$\bar{\omega} = \dot{\alpha}[\bar{i}\sin\phi\cos\beta + \bar{j}\cos\phi\cos\beta + \bar{k}\sin\beta] - \dot{\beta}[\bar{i}\cos\phi - \bar{j}\sin\phi] + \dot{\phi}\bar{k}.$$

$$(3.43)$$

Grouping terms about the unit vectors leads to

$$\bar{\omega} = \bar{i}[\dot{\alpha}\sin\phi\cos\beta - \dot{\beta}\cos\phi] + \bar{j}[\dot{\beta}\sin\phi + \dot{\alpha}\cos\phi\cos\beta]$$
$$+ \bar{k}[\dot{\alpha}\sin\beta + \dot{\phi}].$$

$$(3.44)$$

This equation expresses the rate of rotation vector for the frame as it passes from $OXYZ$ to $Oxyz$ through Euler angles α, β and ϕ in terms of functions of these angles and the unit vectors associated with frame position $Oxyz$. We can use it to express the curvatures of the beam about the x, y and z axes respectively, these are given the notation κ_1, κ_2 and τ respectively. In order to do this we initially decompose the rate of rotation vector for the frame, $\bar{\omega}$, into component parts ω_1, ω_2, ω_3; this is expressed as follows,

$$\bar{\omega} = \bar{i}\omega_1 + \bar{j}\omega_2 + \bar{k}\omega_3.$$

$$(3.45)$$

The rotational velocity components ω_1, ω_2 and ω_3 act about the axes x, y, z respectively, so we can regard them as being functions of the curvature of the beam, κ_1, κ_2, τ such that we have

$$\omega_1 = \kappa_1 \dot{s};$$

$$(3.46)$$

$$\omega_2 = \kappa_2 \dot{s};$$

$$(3.47)$$

$$\omega_3 = \tau \dot{s},$$

$$(3.48)$$

where $\dot{s}$ is the linear velocity along the deformed axis Oz. Substitution of equations (3.46) to (3.48) and (3.45) into equation (3.44) leads to the following curvature relationships,

$$\kappa_1 \dot{s} = \dot{\alpha}\sin\phi\cos\beta - \dot{\beta}\cos\phi,$$

$$(3.49)$$

$$\kappa_2 \dot{s} = \dot{\beta}\sin\phi + \dot{\alpha}\cos\phi\cos\beta,$$

$$(3.50)$$

$$\tau \dot{s} = \dot{\alpha}\sin\beta + \dot{\phi}.$$

$$(3.51)$$

We can transform the independent variables by multiplying each side of each equation by dt/ds which leads to

$$\kappa_1 = \alpha'\sin\phi\cos\beta - \beta'\cos\phi,$$

$$(3.52)$$

$$\kappa_2 = \beta'\sin\phi + \alpha'\cos\phi\cos\beta,$$

$$(3.53)$$

$$\tau = \alpha'\sin\beta + \phi',$$

$$(3.54)$$

where the prime denotes differentiation with respect to s (linear displacement along axis Oz). These latter equations give the three curvatures in terms of functions of the three Euler angles. Of the three Euler angles ϕ is immediately useful, since it is also the twist angle (given in Fig. 2.5(d)), however the relationships between angles α and β and the beam element component displacements u and v are yet to be determined. Once these relationships are clear we will finally be able to express the three important curvature equations exclusively in terms of u, v and ϕ.

The three successive frame rotations of Fig. 3.8 can be related to the actual physical displacements of the structure (as the representative element moves from $OXYZ$ to $Oxyz$), and this is shown diagrammatically in Fig. 3.9. Here, then, is the foundation for this final aspect of the kinematical analysis of the problem.

The most important assumption here is that ds is sufficiently short to be regarded as straight (alternatively that the curvature κ_1 is negligible over such a short length). If we continue to use the prime as an identifier of differentiation with respect to s we will obviously be able to write the following two expressions,

$$du = u' \, ds, \tag{3.55}$$

$$dv = v' \, ds. \tag{3.56}$$

It is also clear from Fig. 3.9 that the following trigonometrical and

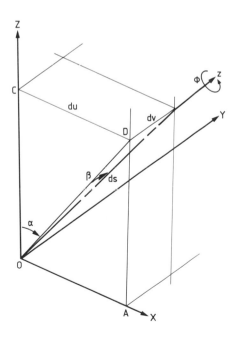

Figure 3.9 Local geometry for combined bending and torsion.

geometrical relationships will hold true,

$$OC = OD \cos \alpha, \tag{3.57}$$

$$(OC)^2 = (OD)^2 - (du)^2, \tag{3.58}$$

$$(OD)^2 = (ds)^2 - (dv)^2. \tag{3.59}$$

Substitution of equations (3.55) and (3.56) into equation (3.58) and (3.59) leads, after some simple manipulation, to

$$\cos \alpha = \left[\frac{1 - (u')^2 - (v')^2}{1 - (v')^2} \right]^{1/2}. \tag{3.60}$$

It is also easily demonstrable that

$$\sin \alpha = \frac{u'}{[1 - (v')^2]^{1/2}}. \tag{3.61}$$

Similar analysis results in the following relationships,

$$\cos \beta = \frac{BD}{ds} = [1 - (v')^2]^{1/2}, \tag{3.62}$$

$$\sin \beta = \frac{dv}{ds} = v'. \tag{3.63}$$

The trigonometrical functions for α and β have now been clarified in terms of u' and v' and what remains is to consider α' and β' so that complete substitutions may be made into equations (3.52)–(3.54).

In order to evlauate α' we divide equation (3.60) by (3.61) so that

$$\tan \alpha = \frac{u'}{[1 - (u')^2 - (v')^2]^{1/2}} = a. \tag{3.64}$$

Therefore

$$\alpha = \tan^{-1} a \tag{3.65}$$

and

$$\alpha' = \frac{1}{(1 + a^2)} a'. \tag{3.66}$$

Differentiating equation (3.64) with respect to s

$$a' = \frac{1}{[1 - (u')^2 - (v')^2]^{1/2}} \frac{u''(1 - (v')^2) + u'v'v''}{(1 - (u')^2 - (v')^2)}. \tag{3.67}$$

Therefore substitution of equations (3.64) and (3.67) into (3.66) gives the result for α'

$$\alpha' = \frac{1}{[1 - (u')^2 - (v')^2]^{1/2}} \left[u'' + \frac{(u'v'v'')}{1 - (v')^2} \right]. \tag{3.68}$$

By a similar procedure we can identify β'

$$\beta' = \frac{v''}{[1 - (v')^2]^{1/2}}. \tag{3.69}$$

Theoretically, at least, we are now in a position to combine equations (3.52)–(3.54), (3.60)–(3.63), (3.68) and (3.69) to obtain the full equations for the curvatures, however these will obviously be cumbersome and unwieldy, although highly accurate. For the sake of practicality a degree of compromise is necessary, and, in order to accomplish this we can choose to neglect certain 'small' terms on the premise that they are of less significance than other 'larger' terms. This concept of 'smallness' and 'largeness' was considered in Section 2.2.3, and in the present example we reiterate the observation that certain derivatives, and certainly quadratic and cubic orders of derivatives, are deemed to be generally of lesser significance than the linear terms. Therefore $v'v''$ and $(v')^2$ are extremely 'small' compared with v (which is itself small), as is $(u')^2$ with u, and so we can write the following approximations,

$$\cos \alpha = 1; \quad \sin \alpha = u'; \quad \cos \beta = 1; \quad \sin \beta = v';$$

$$\alpha' = u''; \quad \beta' = v''. \tag{3.70}$$

Substitution of approximations (3.70) into equations (3.52)–(3.54) results in

$$\kappa_1 \simeq u'' \sin \phi - v'' \cos \phi, \tag{3.71}$$

$$\kappa_2 \simeq v'' \sin \phi + u'' \cos \phi, \tag{3.72}$$

$$\tau \simeq u''v' + \phi'. \tag{3.73}$$

The trigonometrical functions can be eliminated so that the first two equations (3.71) and (3.72) are linear in ϕ by constraining the twist angle such that $\sin \phi \simeq \phi$ and $\cos \phi \simeq 1$. This gives

$$\kappa_1 \simeq u''\phi - v'', \tag{3.74}$$

$$\kappa_2 \simeq v''\phi + u'', \tag{3.75}$$

$$\tau \simeq u''v' + \phi'. \tag{3.76}$$

The beam under analysis here is relatively stiff about axis x because the small displacement v is very much smaller than the beam length l ($v \ll l$); this means that curvature κ_1 is virtually zero. Equation (3.74) then reduces to

$$v'' \simeq u''\phi. \tag{3.77}$$

Substituting for v'' in equations (3.75) and (3.76) enables the curvatures about y and z respectively to be restated

$$\kappa_2 \simeq u''(1 + \phi^2), \tag{3.78}$$

$$\tau \simeq \phi' + u'u''\phi. \tag{3.79}$$

Equation (3.35) allows a transformation of the general lateral displacement $u(z, t)$ to those at the mass centre u_{01}, u_{02} (modal coordinates). This type of representation may also be used for the torsional motion that the system undergoes

$$\phi(z, t) = h(z)\phi_0(t) \tag{3.80}$$

$h(z)$ is a linear mode shape function which, together with f_1 and f_2, may be readily obtained. This equation, along with equations (3.35), (3.78) may be used to define v_0 in terms of u_{01}, u_{02} and ϕ_0, and to this end we need to make one further investigation into the system geometry; this is of the form of Fig. 3.10. In doing this we project the beam curvature on to the u–v plane so that a relationship between u, v and ϕ can be defined. By taking a point P (dz away from O along the deformed z-axis) and then drawing in tangents to these two points on the deformed z-axis one can define a chord $O'P'$ on an end plane as shown in Fig. 3.10(a). This end plane being the area QRS. The distance OQ along the undeformed z-axis is equal to

$$OQ = (l - z) \text{ at } z = \tfrac{1}{2} \text{ say,} \tag{3.81}$$

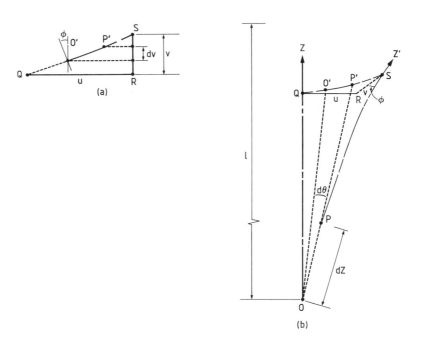

Figure 3.10 Geometrical relationships between u, v, dv, ϕ, dθ and dz for combined bending and torsion.

The chordal length

$$O'P' = (l - z)\,d\theta. \tag{3.82}$$

Figure 3.10(b) shows the relationship between $O'P'$, ϕ and dv, therefore,

$$dv = \phi(O'P'). \tag{3.83}$$

The curvature about the Y-axis is approximately equal to

$$\kappa_2 = \frac{d\theta}{dz}. \tag{3.84}$$

Therefore substitution of equations (3.82) and (3.84) into (3.83) produces the following,

$$dv = (l - z)\phi\kappa_2\,dz. \tag{3.85}$$

Substitution of equation (3.78) into the above and integrating over l/z leads to

$$v = \int_0^{l/z} (l - z)u''(\phi + \phi^3)\,dz. \tag{3.86}$$

The quartic term $u''\phi^3$ is to be neglected because of the difficulties involved in constructing the equations of motion due to the additional higher order non-linearities that would be generated if it was retained. Equations (3.35) and (3.80) are now substituted into equation (3.86), giving,

$$v_0 = \int_0^l (l - z)h\phi_0(f_1''u_{01} + f_2''u_{02})\,dz. \tag{3.87}$$

This can be stated more simply as

$$v_0 = B_1\phi_0 u_{01} + B_2\phi_0 u_{02}, \tag{3.88}$$

where

$$B_1 = \int_0^l (l - z)hf_1''\,dz; \qquad B_2 = \int_0^l (l - z)hf_2''\,dz. \tag{3.89}$$

It is possible to obtain numerical values for B_1 and B_2 which are based on the mode shape functions h, f_1 and f_2 and once these have been ascertained the integration, although somewhat tedious, is not difficult. The results are $B_1 = 0.275$; $B_2 = -0.78$. Since these quantities are determined by the mode shape functions they are not specific to the particular dimensions of the problem under analysis, but relate to a general cantilevered beam with a lumped mass at its tip modelled by means of Euler–Bernoulli beam theory.

(b) System energy considerations

The kinetic energy has been expressed in equation (3.34), and so we need to turn our attention to the potential energy of the system. Timoshenko (1974) gives expressions for strain energy of a beam both in bending and in torsion, these are both contributory to system P.E., and may be written in the following form,

$$U = \tfrac{1}{2}\int_0^l EI_y[u'']^2 \, dz + \tfrac{1}{2}\int_0^l cGJ[\phi']^2 \, dz + m_0 g w_0, \qquad (3.90)$$

where the term $m_0 g w_0$ is the contribution due to gravitational P.E. (through w_0). This term can be related to u by virtue of the expression

$$w = \tfrac{1}{2}\int_0^l \left(\frac{du}{dz}\right)^2 dz = \tfrac{1}{2}\int_0^l (u')^2 \, dz. \qquad (3.91)$$

Therefore the third term of equation (3.90) becomes

$$w_0 = \tfrac{1}{2}\int_0^l (f_1')^2 u_{01}^2 \, dz + \tfrac{1}{2}\int_0^l (f_2')^2 u_{02}^2 \, dz. \qquad (3.92)$$

Equations (3.90) and (3.92) can be put together to give

$$U = \tfrac{1}{2}\left[EI_y\int_0^l (f_1'')^2 \, dz + m_0 g\int_0^l (f_1')^2 \, dz\right]u_{01}^2$$

$$+ \tfrac{1}{2}\left[EI_y\int_0^l (f_2'')^2 \, dz + m_0 g\int_0^l (f_2')^2 \, dz\right]u_{02}^2$$

$$+ \tfrac{1}{2}\left[cGJ\int_0^l (h')^2 \, dz\right]\phi_0^2. \qquad (3.93)$$

Returning to equation (3.34) we can substitute for v_0 by means of equation (3.88) leading to

$$T = \tfrac{1}{2}m_0\left[\dot{u}_{01}^2 + \dot{u}_{02}^2 + \left(\int_0^l (f_1')^2 \, dz\right)u_{01}\dot{u}_{01} + \left(\int_0^l (f_2')^2 \, dz\right)u_{02}\dot{u}_{02}\right.$$

$$\left. + (B_1\dot{\phi}_0 u_{01} + B_1\phi_0\dot{u}_{01} + B_2\dot{\phi}_0 u_{02} + B_2\phi_0\dot{u}_{02} + \dot{W})^2\right] + \tfrac{1}{2}I_0\dot{\phi}_0^2. \qquad (3.94)$$

Applying Lagrange's equations in the form

$$\frac{d}{dt}\frac{\partial T}{\partial \dot{q}_i} - \frac{\partial T}{\partial q_i} + \frac{\partial U}{\partial q_i} = 0, \qquad (3.95)$$

where

$$q_1 = u_{01}; \qquad q_2 = u_{02}; \qquad q_3 = \phi_0$$

enables the governing equations of motion to be written.

In doing this it will become evident that the third and fourth terms ($[\int (f_1')^2 \, dz]u_{01}\dot{u}_{01}$; $[\int (f_2')^2 \, dz]u_{02}\dot{u}_{02}$), disappear because of the Lagrange operations $(d/dt)(\partial T/\partial \dot{u}_{01,02}) - \partial T/\partial u_{01,02}$.

The resulting equations are as follows; note that for convenience subscripts 01 and 02 have been replaced by 1 and 2 for the bending coordinates, and that ϕ_0 becomes ϕ_1

$$\ddot{u}_1(1 + B_1{}^2\phi_1{}^2) + B_1\ddot{\phi}_1\phi_1(B_1u_1 + B_2u_2) + 2B_1\dot{\phi}_1\phi_1(B_1\dot{u}_1 + B_2\dot{u}_2)$$

$$+ B_1B_2\phi_1{}^2\ddot{u}_2 + \omega_1{}^2u_1 + B_1\ddot{W}\phi_1 = 0, \tag{3.96}$$

$$\ddot{u}_2(1 + B_2{}^2\phi_1{}^2) + B_2\ddot{\phi}_1\phi_1(B_1u_1 + B_2u_2) + 2B_2\dot{\phi}_1\phi_1(B_1\dot{u}_1 + B_2\dot{u}_2)$$

$$+ B_1B_2\phi_1{}^2\ddot{u}_1 + \omega_2{}^2u_2 + B_2\ddot{W}\phi_1 = 0, \tag{3.97}$$

$$\ddot{\phi}_1(1 + RB_1{}^2u_1{}^2 + 2RB_1B_2u_1u_2 + RB_2{}^2u_2{}^2)$$

$$+ \dot{\phi}_1(2RB_2{}^2u_2\dot{u}_2 + 2RB_1B_2\dot{u}_1u_2 + 2RB_1B_2u_1\dot{u}_2 + 2RB_1{}^2u_1\dot{u}_1)$$

$$+ \phi_1(RB_1{}^2u_1\ddot{u}_1 + RB_1B_2u_1\ddot{u}_2 + RB_1B_2u_2\ddot{u}_1 + RB_2{}^2u_2\ddot{u}_2)$$

$$+ \omega_3{}^2\phi_1 + R(B_1\ddot{W}u_1 + B_2\ddot{W}u_2) = 0, \tag{3.98}$$

where $R = m_0/I_0$ and ω_1, ω_2, ω_3 are the linear natural frequencies in the fundamental and second bending modes and the fundamental torsion mode respectively. Suitable damping terms in the form of classical linear viscous damping may be included.

Equations (3.96)–(3.98) contain certain linear terms, these are the first one $\ddot{u}_1$, the eighth, $\omega_1{}^2u_1$ and the ninth, $B_1\ddot{W}\phi_1$ (taking equation (3.96) as an example). All the other terms are nonlinear, and contain what are essentially cubic terms. These cubics all arise exclusively from the use of equation (3.88), which defines the form of coupling between u_1, u_2, ϕ_1 and v. Thus it is quite clear that these nonlinear terms are purely geometrical in origin, and (perhaps more specifically), arise because of the orientation of the structure with respect to the imposed base excitation, this orientation in turn admitting the coupling described by equation (3.88) which subsequently generates these terms.

Having identified some of the possible non-linearities which may be present in a structure of this sort (and it should be noted that the simplifications that were highlighted in the foregoing analysis reduced considerably the quantity of non-linearities which were generated) we can progress to an investigation of their effects. Inevitably this is a fairly lengthy operation and involves a perturbation scheme through which the non-linearities may be treated. To accomplish this equations (3.96)–(3.98) need to be ordered by introducing ε in a suitable manner.

Damping in such a system will be relatively low in all three modes with some previous experimental tests revealing that in the case of the two bending modes the numerical values for ξ_b are virtually identical (0.001 nominally, although the exact figure closely depends on beam length l) and the torsional mode damping ratio is even lower at $\xi_t = 0.0005$ nominal. We therefore choose to include damping to $0(\varepsilon^2)$.

Because the significance, or otherwise of the nonlinear terms is unclear at this stage it is prudent to write them to $0(\varepsilon^1)$. The linear terms and the parametric excitation term are of $0(\varepsilon^0)$. Equations (3.96)–(3.98) therefore become

$$\ddot{u}(1 + \varepsilon b_1 B_1 \phi_1{}^2) + 2\varepsilon^2 \zeta_b \omega_1 \dot{u}_1 + \varepsilon b_1 \ddot{\phi}_1 \phi_1 (B_1 u_1 + B_2 u_2)$$
$$+ 2\varepsilon b_1 \dot{\phi}_1 \phi_1 (B_1 \dot{u}_1 + B_2 \dot{u}_2) + \varepsilon b_1 B_2 \phi_1{}^2 \ddot{u}_2$$
$$+ \omega_1{}^2 u_1 + \varepsilon b_1 \ddot{W} \phi_1 = 0, \tag{3.99}$$

$$\ddot{u}_2(1 + \varepsilon b_2 B_2 \phi_1{}^2) + 2\varepsilon^2 \zeta_b \omega_2 \dot{u}_2 + \varepsilon b_2 \ddot{\phi}_1 \phi_1 (B_1 u_1 + B_2 u_2)$$
$$+ 2\varepsilon b_2 \dot{\phi}_1 \phi_1 (B_1 \dot{u}_1 + B_2 \dot{u}_2)$$
$$+ \varepsilon b_1 B_2 \phi_1{}^2 \ddot{u}_1 + \omega_2{}^2 u_2 + \varepsilon b_2 \ddot{W} \phi_1 = 0, \tag{3.100}$$

$$\ddot{\phi}(1 + \varepsilon R b_1 u_1{}^2 + 2\varepsilon R b_1 B_2 u_1 u_2 + \varepsilon R b_2 B_2 u_2{}^2)$$
$$+ \dot{\phi}_1 (2\varepsilon R b_2 B_2 u_2 \dot{u}_2 + 2\varepsilon R b_1 B_2 \dot{u}_1 u_2 + 2\varepsilon R b_1 B_2 u_1 \dot{u}_2$$
$$+ 2\varepsilon R b_1 B_1 u_1 \dot{u}_1) + 2\varepsilon^2 \zeta_t \omega_3 \dot{\phi}_1$$
$$+ \phi_1 (\varepsilon R b_1 B_1 u_1 \ddot{u}_1 + \varepsilon R b_1 B_2 u_1 \ddot{u}_2 + \varepsilon R b_1 B_2 u_2 \ddot{u}_1 + \varepsilon R b_2 B_2 u_2 \ddot{u}_2)$$
$$+ \omega_3{}^2 \phi_1 + \varepsilon R (b_1 \ddot{W} u_1 + b_2 \ddot{W} u_2) = 0, \tag{3.101}$$

where

$$\varepsilon b_1 = B_1; \qquad \varepsilon b_2 = B_2; \qquad \varepsilon^2 \zeta_b = \zeta_b; \qquad \varepsilon^2 \xi_t = \xi_t;$$
$$R = m_0/I_0; \qquad \ddot{W} = -W_0 \Omega^2 \cos \Omega t.$$

We can now proceed to analyse these equations by using the method of multiple scales. The scale of analysis required is fairly extensive and an abridged version which outlines the major steps involved is presented here. A full treatment is available (Cartmell, 1984) with all the necessary intermediate algebra stated in full.

(c) Analysis of the equations of motion

We represent the displacement coordinates u_1, u_2 and ϕ_1 by the following power series, truncated in each case at order ε^2

$$u_1 = u_{10} + \varepsilon u_{11} + \varepsilon^2 u_{12} + \ldots, \tag{3.102}$$

$$u_2 = u_{20} + \varepsilon u_{21} + \varepsilon^2 u_{22} + \ldots, \tag{3.103}$$

$$\phi_1 = \phi_{10} + \varepsilon \phi_{11} + \varepsilon^2 \phi_{12} + \ldots. \tag{3.104}$$

Following the method of Nayfeh (1973) we also express the derivatives in this form to give

$$\frac{d}{dt} = D_0 + \varepsilon D_1 + \varepsilon^2 D_2 + \dots, \tag{3.105}$$

$$\frac{d^2}{dt^2} = D_0{}^2 + 2\varepsilon D_0 D_1 + 2\varepsilon^2 D_0 D_2 + \varepsilon^2 D_1{}^2 + \dots . \tag{3.106}$$

Substitution of the above forms into equations (3.99)–(3.101) and separation of similarly ordered coefficients of ε yields nine perturbation equations (i.e. three to $O(\varepsilon^0)$, three to $O(\varepsilon^1)$, and three to $O(\varepsilon^2)$, for the three equations of motion). The three zeroth-order equations have the following type of solution

$$u_{10} = C_1(T_1, T_2)\exp i\omega_1 T_0 + \bar{C}_1(T_1, T_2)\exp(-i\omega_1 T_0), \tag{3.107}$$

$$u_{20} = C_2(T_1, T_2)\exp i\omega_2 T_0 + \bar{C}_2(T_1, T_2)\exp(-i\omega_2 T_0), \tag{3.108}$$

$$\phi_{10} = C_3(T_1, T_2)\exp i\omega_3 T_0 + \bar{C}_3(T_1, T_2)\exp(-i\omega_3 T_0). \tag{3.109}$$

These solutions can then be substituted into the first-order perturbation equation so that solutions to these can be found. In order to do this properly we find that in each first-order perturbation equation there are terms that must be removed in order to ensure that simple first-order resonances are avoided (meaning those resonances involving $\Omega = \omega_1$ and $\Omega = \omega_2$ and $\Omega = \omega_3$). Therefore the following conditions are imposed

$$i2D_1 C_j \omega_j \exp i\omega_j T_0 + \text{C.C.} = 0; \qquad j = 1, 2, 3, \tag{3.110}$$

where C.C. denotes the complex conjugate of the preceding term. Equating these terms to zero now allows us to proceed with the solution of the first-order perturbation equations without the resonances outlined above confusing the subsequent analysis.

Equations (3.110) show that $D_1 C_j = 0$; $j = 1, 2, 3$, meaning that the C_j are functions only of time scale T_2 and not T_1. Solutions to the first-order perturbation equation can now be sought, and turn out to be of the following form

$$u_{11} = \sum_{k=1}^{6} K_k \exp i(p_k\Omega \pm q_k\omega_1 \pm r_k\omega_2 \pm s_k\omega_3) + \text{C.C.}, \tag{3.111}$$

$$u_{21} = \sum_{m=1}^{6} K_m \exp i(p_m\Omega \pm q_m\omega_1 \pm r_m\omega_2 \pm s_m\omega_3) + \text{C.C.}, \tag{3.112}$$

$$\phi_{11} = \sum_{n=1}^{12} K_n \exp i(p_n\Omega \pm q_n\omega_1 \pm r_n\omega_2 \pm s_n\omega_3) + \text{C.C.}, \tag{3.113}$$

where the $p_{k,m,n}$, $q_{k,m,n}$, $r_{k,m,n}$, $s_{k,m,n}$ are integer coefficients (or zero), and the $K_{k,m,n}$ are complicated amplitude and frequency dependent coefficients (which do not need to be explicitly stated at this stage of the analysis in order to follow the procedures involved in the expansion).

The zeroth-order perturbation solutions (equations (3.107)–(3.109)) and the first-order perturbation solutions (equations (3.111)–(3.113)) may now be substituted, as required, into the second-order perturbation equations (those to $0(\varepsilon^2)$) so that we arrive at the following,

$$D_0^2 u_{12} + \omega_1^2 u_{12}$$

$$= \exp i\omega_1 T_0 \left[\sum_{j=1}^{26} L_j \exp i(a_j \Omega + b_j \omega_1 + d_j \omega_2 + e_j \omega_3) T_2 + \text{C.C.} \right],$$

$$(3.114)$$

$$D_0^2 u_{22} + \omega_2^2 u_{22}$$

$$= \exp i\omega_2 T_0 \left[\sum_{j=27}^{51} L_j \exp i(a_j \Omega + b_j \omega_1 + d_j \omega_2 + e_j \omega_3) T_2 + \text{C.C.} \right],$$

$$(3.115)$$

$$D_0^2 \phi_{12} + \omega_3^2 \phi_{12}$$

$$= \exp i\omega_3 T_0 \left[\sum_{j=52}^{78} L_j \exp i(a_j \Omega + b_j \omega_1 + d_j \omega_2 + e_j \omega_3) T_2 + \text{C.C.} \right]$$

$$(3.116)$$

The a_j, b_j, d_j, e_j are integer coefficients (or zero dependent on the particular term) and the L_j are coefficients which contain system constants and response amplitudes. The exponent of each L_j term is potentially capable of indicating the response condition (which on substitution into the exponent will reduce that term to a static term rather than one which grows with time-scale T_2). For example if we take the hypothetical case in which (in equation (3.114)), $j = m$ and $a_m = 1$, $b_m = -2$, $d_m = -1$, $e_m = 1$ then the mth term becomes, $L_{j=m} \exp i(\Omega - 2\omega_1 - \omega_2 + \omega_3) T_2$; and if we then substitute the resonance $\Omega = 2\omega_1 + \omega_2 - \omega_3$ we then get $L_{j=m} \exp i(0) T_2 = L_{j=m}$. Thus this term as interpreted above is resonant to $\Omega = 2\omega_1 + \omega_2 - \omega_3$ and will contribute to a solution which is resonant to $\Omega = 2\omega_1 + \omega_2 - \omega_3$ (without u_{12} growing too large and invalidating the uniformity of the expansion of equation (3.102)), if we remove it (and others, as appropriate to the resonance(s) of interest) and subsequently equate to zero. This, of course, is the standard multiple scale procedure for treating such so-called secular terms.

All the right-hand side terms admit resonances of one sort or another; a full list of these is given in Table 3.1 in which it may be seen that some appear in one, two, or all three of the second-order perturbation equations. We are only interested in those resonances that show coupling between at least two modes as these are combination resonances (which derive from the kinematical coupling of equation (3.88),

Table 3.1

Equation (3.81)	Equation (3.82)	Equation (3.83)
$\Omega = \omega_1 - 2\omega_2 - \omega_3$		$\Omega = \omega_1 - 2\omega_2 - \omega_3$
$\Omega = \omega_1 - 2\omega_2 + \omega_3$		
$\Omega = -\omega_2 - \omega_3$	$\Omega = -\omega_2 - \omega_3$	$\Omega = -\omega_2 - \omega_3$
$\Omega = 2\omega_1 - \omega_2 - \omega_3$		
$\Omega = -\omega_2 + \omega_3$	$\Omega = -\omega_2 + \omega_3$	$\Omega = -\omega_2 + \omega_3$
$\Omega = 2\omega_1 - \omega_2 + \omega_3$		$\Omega = 2\omega_1 - \omega_2 + \omega_3$
$\Omega = -\omega_1 - \omega_3$	$\Omega = -\omega_1 - \omega_3$	$\Omega = -\omega_1 - \omega_3$
$\Omega = 3\omega_1 - \omega_3$		
$\Omega = \omega_1$		
$\Omega = \frac{1}{2}(\omega_1 - \omega_2)$		
$\Omega = \frac{1}{2}(\omega_1 + \omega_2)$	$\Omega = \frac{1}{2}(\omega_1 + \omega_2)$	
$\Omega = \omega_1 - \omega_3$	$\Omega = \omega_1 - \omega_3$	$\Omega = \omega_1 - \omega_3$
$\Omega = \omega_1 + \omega_3$	$\Omega = \omega_1 + \omega_3$	$\Omega = \omega_1 + \omega_3$
$\Omega = \omega_2 - \omega_3$	$\Omega = \omega_2 - \omega_3$	$\Omega = \omega_2 - \omega_3$
$\Omega = 2\omega_1 + \omega_2 - \omega_3$	$\Omega = 2\omega_1 + \omega_2 - \omega_3$	
$\Omega = -\omega_1 + \omega_3$	$\Omega = -\omega_1 + \omega_3$	$\Omega = -\omega_1 + \omega_3$
$\Omega = \omega_2 + \omega_3$	$\Omega = \omega_2 + \omega_3$	$\Omega = \omega_2 + \omega_3$
$\Omega = 3\omega_1 + \omega_3$		$\Omega = 3\omega_1 + \omega_3$
$\Omega = 2\omega_1 + \omega_2 + \omega_3$	$\Omega = 2\omega_1 + \omega_2 + \omega_3$	$\Omega = 2\omega_1 + \omega_2 + \omega_3$
$\Omega = \omega_1 - 3\omega_3$		
$\Omega = \omega_1 + 3\omega_3$		$\Omega = \omega_1 + 3\omega_3$
$\Omega = \omega_1 + 2\omega_2 - \omega_3$	$\Omega = \omega_1 + 2\omega_2 - \omega_3$	
$\Omega = \omega_1 + 2\omega_2 + \omega_3$	$\Omega = \omega_1 + 2\omega_2 + \omega_3$	$\Omega = \omega_1 + 2\omega_2 + \omega_3$
	$\Omega = -2\omega_1 + \omega_2 - \omega_3$	
	$\Omega = \frac{1}{2}(\omega_2 - \omega_1)$	
	$\Omega = \omega_2$	
	$\Omega = 2\omega_1 + \omega_2 - \omega_3$	
	$\Omega = 3\omega_2 + \omega_3$	$\Omega = 3\omega_2 + \omega_3$
	$\Omega = 3\omega_2 - \omega_3$	
	$\Omega = \omega_2 - 3\omega_3$	
	$\Omega = \omega_2 + 3\omega_3$	$\Omega = \omega_2 + 3\omega_3$
	$\Omega = -\omega_1 + 2\omega_2 + \omega_3$	$\Omega = -\omega_1 + 2\omega_2 + \omega_3$
	$\Omega = -2\omega_1 + \omega_2 + \omega_3$	$\Omega = -2\omega_1 + \omega_2 + \omega_3$
		$\Omega = -\omega_1 + 3\omega_3$
		$\Omega = -\omega_1 - 2\omega_2 + \omega_3$
		$\Omega = -3\omega_2 + \omega_3$
		$\Omega = -2\omega_1 - \omega_2 + \omega_3$
		$\Omega = -3\omega_1 + \omega_3$
		$\Omega = \omega_3$

which in turn was the source of the non-linearities present in equations (3.99)–(3.101)).

Section 2.4.1(c) considers one resonance from the above table (refer to equation (2.204)) and it goes on to state that an expression for the detuning parameter $\varepsilon^2 \rho_1$ can be derived (after some considerable algebraic manipulation (Cartmell, 1984; Cartmell and Roberts 1987)).

Figures 2.11 and 2.12 show the instability zoning and the modal responses with time respectively. For the sake of completeness we will restate the resonance expression (equation (2.204)) and the detuning parameter expression (equation (2.205))

$$\Omega = \tfrac{1}{2}(\omega_1 + \omega_2) + \varepsilon^2 \rho_1; \qquad V_1(\varepsilon^2 \rho_1)^2 + V_2(\varepsilon^2 \rho_1) + V_3 = 0,$$

where the coefficients V_1, V_2 and V_3 are as follows

$$V_1 = \left[\frac{64 q_1 r_1 \omega_1^2}{R^2 B_1^2 (W_0 \Omega^2)^4} + \frac{16 p_2 q_1 \omega_1}{R(W_0 \Omega^2)^2} + q_2 - \frac{B_2^2 r_3 (\omega_1 + \omega_2)^2}{\omega_2^2} \right.$$
$$\left. + \frac{64 \xi_b^2 \omega_1^2 (\omega_1 + \omega_2)^2 (r_1 q_2 + r_3 q_1)}{R^2 B_1^2 (W_0 \Omega^2)^4} \right], \tag{3.117}$$

$$V_2 = \left[\frac{16 p_1 q_1 \omega_1}{R(W_0 \Omega^2)^2} - \frac{64 \xi_b^2 \omega_1^2 r_2 q_1 (\omega_1 + \omega_2)^2}{R^2 B_1^2 (W_0 \Omega^2)^4} \right.$$
$$\left. - \frac{B_2^2 r_2 (\omega_1 + \omega_2)^2}{\omega_2^2} \right], \tag{3.118}$$

$$V_3 = \left[B_1^2 q_1 + \frac{64 \xi_b^2 \omega_1^2 r_1 q_1 (\omega_1 + \omega_2)^2}{R^2 B_1^2 (W_0 \Omega^2)^4} \right.$$
$$\left. - \frac{B_2^2 r_1 (\omega_1 + \omega_2)^2}{\omega_2^2} \right], \tag{3.119}$$

with

$$p_1 = [\omega_3^2 - \tfrac{9}{4}\omega_1^2 - \tfrac{1}{4}\omega_2^2 - \tfrac{3}{2}\omega_1 \omega_2] \tag{3.120}$$

$$p_2 = -[3\omega_1 + \omega_2] \tag{3.121}$$

$$q_1 = [\omega_3^4 - \tfrac{1}{2}\omega_3^2 \omega_2^2 - \tfrac{1}{2}\omega_3^2 \omega_1^2 + \omega_1 \omega_2 \omega_3^2 + \tfrac{9}{16}\omega_1^2 \omega_2^2$$
$$+ \tfrac{1}{16}\omega_1^4 - \tfrac{1}{4}\omega_1^3 \omega_2 + \tfrac{1}{16}\omega_2^4 - \tfrac{1}{4}\omega_1 \omega_2^3] \tag{3.122}$$

$$q_2 = [\omega_1 \omega_2 - 2\omega_3^2 - \tfrac{1}{2}\omega_1^2 - \tfrac{1}{2}\omega_2^2] \tag{3.123}$$

$$r_1 = [\omega_3^4 - \tfrac{9}{2}\omega_1^2 \omega_3^2 - \tfrac{1}{2}\omega_2^2 \omega_3^2 - 3\omega_1 \omega_2 \omega_3^2 + \tfrac{81}{16}\omega_1^4$$
$$+ \tfrac{27}{8}\omega_1^2 \omega_2^2 + \tfrac{27}{4}\omega_1^3 \omega_2 + \tfrac{3}{4}\omega_1 \omega_2^3 + \tfrac{1}{16}\omega_2^4] \tag{3.124}$$

$$r_2 = [\tfrac{27}{2}\omega_1^3 + \tfrac{27}{2}\omega_1^2 \omega_2 + \tfrac{9}{2}\omega_1 \omega_2^2 + \tfrac{1}{2}\omega_2^3 - 6\omega_1 \omega_3^2 - 2\omega_2 \omega_3^2] \tag{3.125}$$

$$r_3 = [\tfrac{27}{2}\omega_1^2 + \tfrac{3}{2}\omega_2^2 + 9\omega_1 \omega_2 - 2\omega_3^2]. \tag{3.126}$$

It is not an easy exercise to trace the source of each of the combinations given in Table 3.1, however it is true to say that the nonlinear terms in equations (3.99)–(3.101) are the general source of most of these combination resonances. The nonlinear terms are all essentially cubic in form with coordinate coupling exhibited in every one. The linear part of these equations, if written separately, constitute a three degree of freedom, linear parametric problem with coupling in the parametric excitation terms

$$\ddot{u}_1 + 2\varepsilon^2\zeta_b\omega_1\dot{u}_1 + \omega_1{}^2 u_1 + \varepsilon b_1\ddot{W}\phi_1 = 0, \tag{3.127}$$

$$\ddot{u}_2 + 2\varepsilon^2\zeta_b\omega_2\dot{u}_2 + \omega_2{}^2 u_2 + \varepsilon b_2\ddot{W}\phi_1 = 0, \tag{3.128}$$

$$\ddot{\phi}_1 + 2\varepsilon^2\zeta_t\omega_t\dot{\phi}_1 + \omega_3{}^2\phi_1 + \varepsilon R(b_1\ddot{W}u_1 + b_2\ddot{W}u_2) = 0 \tag{3.129}$$

(these are identical in form to equations (2.201)–(2.203)).

Reduced equations of this sort will lay claim to resonances such as $\Omega - \omega_1 + \omega_3$, $\Omega - \omega_2 + \omega_3$, $\Omega - \omega_2 - \omega_3$, $\Omega - \frac{1}{2}(\omega_1 + \omega_2)$. The external resonances $\Omega = \omega_1$, ω_2, ω_3 are relatively trivial and in the foregoing analysis can be shown to be of second order status (having been eliminated to first order by the removal of secular terms (equation (3.110)). Second-order status for all the other terms of Table 3.1 (that is, those which were primarily generated by the non-linear parts of equations (3.99)–(3.101)) does not imply insignificance though, since the first-order perturbation solutions fundamentally underpin them (Cartmell, 1984; Cartmell and Roberts, 1987).

3.2 MATERIAL AND STRUCTURAL CONFIGURATION NON-LINEARITIES

3.2.1 Power law damping in a parametrically excited system

Nonlinear damping as experienced by a body moving through a fluid is generally thought to be fairly accurately expressed by a velocity squared term such as $c\dot{x}^2$ (or $c\dot{x}|\dot{x}|$); this is sometimes known as quadratic damping.

Other forms of nonlinear damping arising from the material structure of certain alloys and composites can also be represented by some sort of power law; $c\dot{x}^v$, where v is not necessarily an integer. We will take the quadratic case and investigate its effects on a parametrically excited nonlinear oscillator, noting that this quadratic form applies principally to high Reynolds number problems (i.e. those in which the velocities are high and/or the liquid viscosity is low). The reader is referred to a general discussion on damping mechanisms in Section 1.2.3 and the references quoted there for further details on nonlinear damping mechanisms.

Quadratic damping necessitates an approximative solution technique such as multiple scales (Nayfeh and Mook, 1979) or alternatively an averaging method such as Krylov–Bogoliobov–Mitropolsky (K–B–M) (Hsu, 1975a). The latter approach has successfully been used in the treatment of the current problem (Hsu, 1975b); a review of this is to be given here. The fundamental equation for a nonlinear (cubic) oscillator with one degree of freedom and a linear parametric excitation term is as follows (note that the nonlinear damping term is left in generalized form, to order α, for now)

$$\ddot{x} + c_1\dot{x} + c_\alpha|\dot{x}|^\alpha \operatorname{sgn} x + (\delta + \varepsilon \cos t)(x + \beta x^3) = 0. \quad (3.130)$$

It is not proposed that a description of the K–B–M method should be given here since it is considered that multiple scales would give a comparable result, however such details may be obtained by recourse to the original work (Bogoliobov and Mitropolsky, 1961).

Application of the K–B–M technique to this problem yields two first-order differential equations

$$\dot{a} = -c_1 + \varepsilon \sin 2\theta - c_\alpha b_1(a/2)^{\alpha-1}, \quad (3.131)$$

$$\dot{\theta} = 2\gamma + \varepsilon \cos 2\theta + \tfrac{3}{8}\beta a^2, \quad (3.132)$$

where α and θ are the amplitude and phase constituents of the principal part of the assumed solution, thus,

$$x = a \cos((n/2)t + \theta) + \text{higher order terms.} \quad (3.133)$$

The other variables and constants of equations (3.131) and (3.132) are defined as

c_1, linear damping coefficient
c_α, nonlinear damping coefficient, where α = order of damping
$\delta = (n/2)^2 + \gamma$, natural frequency and detuning parameter, γ
ε, parametric excitation magnitude
β, cubic non-linearity magnitude coefficient
b_1, Fourier coefficient.

Since this is a nonlinear problem we are looking at the stablilizing effects of the non-linearities on the otherwise unbounded linear solutions. To this end we require steady-state (periodic) solutions, which, in the original work, are designated by a^* and θ^* for amplitude and phase respectively. So in order to get steady-state solutions we set $\dot{a} = \dot{\theta} = 0$ in equations (3.131) and (3.132). The trivial solutions ($a^* = 0$ and $\cos 2\theta^* = -2\gamma/\varepsilon$) are of little interest, with the non-trivial ($a^* \neq 0$) solutions remaining; these can be found by eliminating θ^* from equations (3.131) and (3.132) (after having set $\dot{a} = \dot{\theta} = 0$), this gives the following

$$[2\gamma + \tfrac{3}{8}\beta(a^*)^2]^2 + (c_1 + c_\alpha(a^*/2)^{\alpha-1}b_1)^2 - \varepsilon^2 = 0. \quad (3.134)$$

The linear undamped problem is given by $c_1 = c_\alpha = \beta = 0$ for which equations (3.134) has no non-trivial stationary solutions. The linear, damped, problem (which is defined by $c_\alpha = \beta = 0$) also has no non-trivial stationary solutions.

In both these cases unbounded solutions will occur within specific $\varepsilon-\gamma$ domains $(-|\varepsilon|/2 \leqslant \gamma \leqslant |\varepsilon|/2)$ for the undamped case, and $-(\varepsilon^2 - c_1^2)^{1/2}/2 \leqslant \gamma \leqslant (\varepsilon^2 - c_1^2)^{1/2}/2$ for the case of linear damping, $c_1 \neq 0$.

The case for which linear damping and the cubic are zero, but the nonlinear damping is quadratic, such that,

$$c_\alpha = c_2; \qquad b_m = b_1 = -\left(\frac{8}{\pi m(m^2 - 4)}\right) = \frac{8}{3\pi}; \qquad c_1 = \beta = 0,$$

leads to a non-trivial stationary response given by

$$a^* = \frac{3\pi}{4c_2} \sqrt{\varepsilon^2 - 4\gamma^2}. \qquad (3.135)$$

If we now make the linear damping non-zero, $c_1 \neq 0$, then this stationary response equation extends to

$$a^* = \frac{3\pi}{4c_2} [\sqrt{\varepsilon^2 - 4\gamma^2} - c_1]. \qquad (3.136)$$

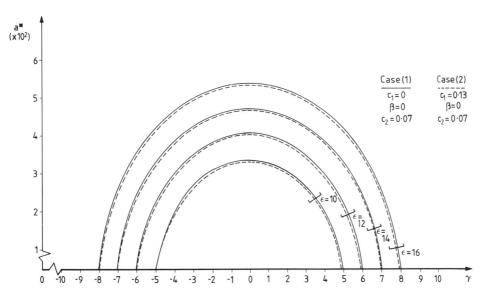

Figure 3.11 Theoretical response of a single degree of freedom system with linear and quadratic damping, linear stiffness and linear parametric excitation; coefficient of cubic term is zero in both quoted cases (i.e. $\beta = 0$) (data and response curves derived by the author with reference to the work of Hsu, 1975b).

The general case, $c_1 \neq 0$, $c_2 \neq 0$; $\beta \neq 0$ is given in equation (3.134) ($\alpha = 2$).

The three preceding stationary response cases are shown in Figs 3.11 (first two cases) and 3.12 (third case).

The curves of Fig. 3.11 (unbroken lines) represent the first case, in which the only non-linearity is the quadratic damping function. Linear viscous damping is set to zero ($c_1 = 0$). The curves show that the nonlinear damping serves to bound the responses. Effectively the system is stabilized over a range of detuning values, γ. As one would expect this region and the maximum amplitude, (at $\gamma = 0$), both increase with increasing parametric excitation magnitude, ε.

The second case, which is shown as broken line on the same figure, gives responses for the conditions above plus non-zero linear damping. Even though the case shown here is for a linear damping coefficient, which is almost twice that of the quadratic term, its overall effect is minimal, reducing the response amplitudes by an amount equal to $3\pi c_1/4c_2$ (refer to equation (3.136)).

It is important to recall that the purely linear parametric problem,

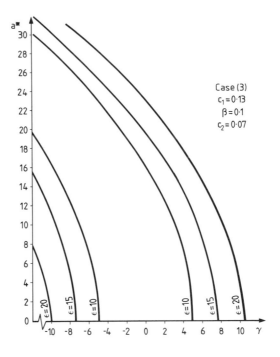

Figure 3.12 Effect of including the cubic non-linearity in the system of equation (3.130) (data and response curves derived by the author with reference to the work of Hsu, 1975a).

with linear damping, shows us that the magnitude of the damping term does not in any way affect the magnitude of the responses. These grow in an unbounded manner with time. What it does do, however, is to pull the threshold point of the unstable region back away from the frequency axis. In this present nonlinear problem we also find that the linear damping term is of little significance with regard to the response magnitudes, even through these are finite and bounded.

Finally, the third case, for which $\beta \neq 0$, is shown in Fig. 3.12 and we see how important this term is in its modifying effects. The curves still widen out over the γ axis for increasing ε, however there is a very marked skew (the extent of which is governed by the magnitude of β). The global response magnitudes are dramatically reduced and (although it is not shown here) we can easily demonstrate how inconsequential the actual magnitudes of c_1 and c_2 are now the cubic non-linearity is incorporated. Variation of these over a fairly large range has very little effect on a^*, or on the width of the detuning region.

3.2.2 Physical configuration non-linearities

(a) Quadratic configuration non-linearities in parametric systems

We have already examined the case of quadratic damping (Section 3.2.1) and can now move on to quadratic terms which result because of other physical effects. These are numerous and diverse, covering such cases as rotating shaft motion, arch vibration, shell vibration, structural oscillations about a loaded static equilibrium point, pitch and roll motion of ships and elastically supported liquid sloshing problems to name but a few. Since most problems of practical interest involve many participating and interacting modes (and therefore degrees of freedom), we will consider a multi-degree-of-freedom representation of a general problem which embraces generalized quadratic non-linearities which could describe the various examples cited above. In order to accomplish this we make references to the work of Nayfeh (1983). It is perhaps an apt point at which to mention the very large and original contribution made by Nayfeh to the whole field of parametric and nonlinear vibration studies over recent years. Once again, it is highly recommended that his definitive text be investigated for a thorough understanding of problems and ideas that can only be cursorily touched on here (Nayfeh and Mook, 1979). Returning to the problem in hand, Nayfeh proposes the following general equations for the study of quadratic non-linearities

$$\ddot{u}_n + \omega_n{}^2 u_n + 2\varepsilon\mu_n\dot{u}_n - 2\varepsilon \sum_{m,k}^{N} \alpha_{nmk}u_m u_k$$

$$- 2\varepsilon \sum_{m}^{N} f_{nm} \cos \Omega t . u_m = 0, \qquad (\alpha_{nmk} \equiv \alpha_{nkm}), (3.137)$$

where ε is a small parameter and can be considered to be identical in effect to a perturbation parameter. The α_{nmk} and f_{nm} are quadratic non-linearity magnitudes and parametric excitation term magnitudes (i.e. constants) respectively. The modal frequencies ω_1 to ω_N are such that $\omega_1 < \omega_2 < \omega_3 < \omega_4 < \ldots < \omega_N$. These equations may be solved by recourse to the method of multiple scales and a first-order expansion for u_n and the first and second derivatives.

$$u_n = u_{n0}(T_0, T_1) + \varepsilon u_{n1}(T_0, T_1) + \ldots, \qquad (3.138)$$

$$\frac{\mathrm{d}}{\mathrm{d}t} = D_0 + \varepsilon D_1 + \ldots, \qquad (3.139)$$

$$\frac{\mathrm{d}^2}{\mathrm{d}t^2} = D_0{}^2 + 2\varepsilon D_0 D_1 + \ldots. \qquad (3.140)$$

Nayfeh follows his usual method of analysis from here and substitutes equations (3.138)–(3.140) into equations (3.137) leading to the zeroth- and first-order perturbation equations. The solutions to the zeroth-order equations are stated in the conventional complex form

$$u_{n0} = A_n \exp(i\omega_n T_0) + \bar{A}_n \exp(-i\omega_n T_0). \qquad (3.141)$$

These can now be appropriately substituted into the first-order perturbation equations to give

$$\begin{aligned}
D_0{}^2 u_{n1} + \omega_n{}^2 u_{n1} = \exp(i\omega_n T_0)\{&-i2\omega_n(D_1 A_n + \mu_n A_n) \\
&+ 2\sum_{m,k}^{N} \alpha_{nmk}[A_m A_k \exp i(\omega_m + \omega_k - \omega_n)T_0 \\
&+ A_m \bar{A}_k \exp i(\omega_m - \omega_k - \omega_n)T_0] \\
&+ \sum_{m}^{N} f_{nm}[A_m \exp i(\Omega + \omega_m - \omega_n)T_0 \\
&+ \bar{A}_m \exp i(\Omega - \omega_m - \omega_n)T_0]\} + \text{C.C.} \qquad (3.142)
\end{aligned}$$

As is usual with multiple scales we will find that any particular solutions to equations such as (3.142) will contain so-called secular terms due to the presence of internal resonances and certain parametric resonances. We will illustrate potential effects by considering interaction between three modes such that the following resonances apply

$$\omega_3 = \omega_1 + \omega_2 + \varepsilon\sigma, \qquad (3.143)$$

$$\Omega = 2\omega_1 + \varepsilon\sigma_1, \qquad (3.144)$$

$$\Omega = 2\omega_3 + \varepsilon\sigma_3. \qquad (3.145)$$

We can now substitute these resonances into the exponents of the right-hand side terms of equations (3.142) and can remove any that appear to be capable of generating secular terms. We describe this operation as setting up the solvability conditions. So for $n = 1$ we get the following by making judicious choices for m and k, noting that for three modes (and $n = 1$), $m = 1$, 2 or 3, $k = 1$, 2 or 3,

$$-\mathrm{i}2\omega_1(D_1A_1 + \mu_1A_1) + 2(\alpha_{123} + \alpha_{132})A_3\bar{A}_2 \exp\mathrm{i}(\varepsilon\sigma)T_0$$

$$+ e_1 f_{11}\bar{A}_1 \exp\mathrm{i}(\varepsilon\sigma T_0) = 0. \tag{3.146}$$

For $n = 2$; $m = 1$, 2 or 3; $k = 1$, 2 or 3 the following solvability conditions emerge

$$-\mathrm{i}2\omega_2(D_1A_2 + \mu_2A_2) + 2(\alpha_{231} + \alpha_{213})A_3\bar{A}_1 \exp\mathrm{i}(\varepsilon\sigma T_0) = 0. \tag{3.147}$$

Finally for $n = 3$ we find

$$-\mathrm{i}2\omega_3(D_1A_3 + \mu_3A_3) + 2(\alpha_{312} + \alpha_{321})A_2A_1 \exp\mathrm{i}(-\varepsilon\sigma T_0)$$

$$+ e_3 f_{33}\bar{A}_3 \exp\mathrm{i}(\varepsilon\sigma T_0) = 0, \tag{3.148}$$

where $\alpha_{123} = \alpha_{132}$; $\alpha_{231} = \alpha_{213}$; $\alpha_{312} = \alpha_{321}$ and also where $e_{1,3}$ are either unity or zero such that

$$e_1 = 1; \qquad e_3 = 0 \qquad \text{for} \qquad \Omega = 2\omega_1 + \varepsilon\sigma_1, \tag{3.149}$$

$$e_1 = 0; \qquad e_3 = 1 \qquad \text{for} \qquad \Omega = 2\omega_3 + \varepsilon\sigma_3. \tag{3.150}$$

Since resonances (3.143)–(3.145) do not include modal frequencies for $n \geq 4$ then the following occurs for $n \geq 4$,

$$D_1A_n + \mu_nA_n = 0. \tag{3.151}$$

Therefore the A_n decays to zero with time-scale T_1 (and hence t); also we see that since modes higher than the third do not appear in equations (3.146)–(3.148) we can correctly restrict further analysis to these equations. The complex amplitudes may be expressed in this form

$$A_{1,2,3} = \tfrac{1}{2}a_{1,2,3} \exp(\mathrm{i}\beta_{1,2,3}), \tag{3.152}$$

(where $a_{1,2,3} = a_{1,2,3}(T_1)$; $\beta_{1,2,3} = \beta_{1,2,3}(T_1)$).

Differentiating the A_n with respect to time scale T_1 as required, and separating out the real and imaginary parts produces a set of six first-order slow time equations, noting the use of the prime to denote differentiation with respect to time-scale, T_1,

$$a_1' = -\mu_1a_1 + \frac{\alpha_{123}}{\omega_1}a_3a_2 \sin(\beta_3 - \beta_2 - \beta_1 + \varepsilon\sigma T_0)$$

$$+ e_1\frac{f_{11}}{2\omega_1}a_1 \sin(\varepsilon\sigma_1 T_0 - 2\beta_1), \tag{3.153}$$

$$\beta_1' = - \frac{\alpha_{123}}{\omega_1} \frac{a_3 a_2}{a_1} \cos{(\beta_3 - \beta_2 - \beta_1 + \varepsilon\sigma T_0)}$$

$$- e_1 \frac{f_{11}}{2\omega_1} a_1 \cos{(\varepsilon\sigma_1 T_0 - 2\beta_1)}, \tag{3.154}$$

$$a_2' = - \mu_2 a_2 + \frac{\alpha_{231}}{\omega_2} a_1 a_3 \sin{(\beta_3 - \beta_2 - \beta_1 + \varepsilon\sigma T_0)}, \tag{3.155}$$

$$\beta_2' = - \frac{\alpha_{231}}{\omega_2} \frac{a_1 a_3}{a_2} \cos{(\beta_3 - \beta_2 - \beta_1 + \varepsilon\sigma T_0)}, \tag{3.156}$$

$$a_3' = - \mu_3 a_3 - \frac{\alpha_{312}}{\omega_3} a_1 a_2 \sin{(\beta_3 - \beta_2 - \beta_1 + \varepsilon\sigma T_0)}$$

$$+ e_3 \frac{f_{33}}{2\omega_3} a_3 \sin{(\varepsilon\sigma_3 T_0 - 2\beta_3)}, \tag{3.157}$$

$$\beta_3' = - \frac{\alpha_{312}}{\omega_3} \frac{a_1 a_2}{a_3} \cos{(\beta_3 - \beta_2 - \beta_1 + \varepsilon\sigma T_0)}$$

$$- e_3 \frac{f_{33}}{2\omega_3} a_3 \cos{(\varepsilon\sigma_3 T_0 - 2\beta_3)}. \tag{3.158}$$

The trigonometrical function arguments could be defined as follows,

$$\gamma = \beta_3 - \beta_2 - \beta_1 + \varepsilon\sigma T_0 = \beta_3 - \beta_2 - \beta_1 + \sigma T_1, \tag{3.159}$$

$$\gamma_1 = \varepsilon\sigma_1 T_0 - 2\beta_1 = \sigma_1 T_1 - 2\beta_1, \tag{3.160}$$

$$\gamma_3 = \varepsilon\sigma_3 T_0 - 2\beta_3 = \sigma_3 T_1 - 2\beta_3, \tag{3.161}$$

with the angles γ, γ_1, γ_3, being known as autonomous system phase angles. For a steady state system we impose conditions such that

$$a_n' = \gamma' = \gamma_1' = \gamma_3' = 0. \tag{3.162}$$

Nayfeh describes the external resonance condition

$$\Omega = 2\omega_1 + \varepsilon\sigma_1, \quad \Omega \text{ away from } 2\omega_3 + \varepsilon\sigma_3$$

as a fundamental (parametric) resonance of the first mode, thus $e_1 = 1$; $e_3 = 0$, and the internal intermodal coupling of equation (3.143) still holds. Returning to equations (3.159)–(3.161) we find

$$\beta_3' - \beta_2' - \beta_1' + \sigma = 0 \tag{3.163}$$

and

$$\beta_1' = \frac{\sigma_1}{2} \tag{3.164}$$

leading to

$$\beta_3' - \beta_2' = \frac{\sigma_1}{2} - \sigma. \tag{3.165}$$

Equation (3.165) may now be used in conjunction with equations (3.156) and (3.158) to eliminate B_n'. On examination of the resulting equations for steady state it becomes apparent that $a_n = 0$ is possible. These are mathematically correct but uninteresting, trivial, solutions. Of more significance is the case where $a_n \neq 0$, and for this case it can easily be shown that

$$\frac{a_3{}^2}{a_2{}^2} = -\frac{\mu_2 \omega_2 \alpha_{312}}{\mu_3 \omega_3 \alpha_{231}}. \tag{3.166}$$

Unfortunately this has no physical significance because damping μ_n and natural frequencies ω_n are positive in this problem. Similarly α_{123}, α_{231} and α_{321} are all of the same sign. Therefore for $\Omega = 2\omega_1 + \varepsilon \sigma_1$ the system only admits trivial zero solutions.

Fortunately this rather anticlimactic state of affairs is not the case for the other fundamental parametric resonance possibility, upon which this analysis has been based, namely $\Omega - 2\omega_3 + \varepsilon \sigma_3$. Obviously now $e_1 = 0$ and $e_3 = 1$, and equation (3.161) gives us the following

$$\beta_3' = \frac{\sigma_3}{2}. \tag{3.167}$$

Also, from equations (3.163) and (3.167) we get

$$\beta_2' + \beta_1' = \sigma + \frac{\sigma_3}{2}. \tag{3.168}$$

Setting $a_n' = 0$ and introducing equations (3.167) and (3.168) into the slow time equations (3.153) to (3.158) enables these two emerging cases to be investigated; firstly $a_n = 0$ if

$$f_{33}{}^2 \leq \omega_3{}^2(\sigma_3{}^2 + 4\mu_3{}^2) \tag{3.169}$$

and secondly $a_n \neq 0$ for which a certain amount of algebraic manipulation is required before the solutions are identified

$$\frac{a_1{}^2}{a_2{}^2} = \frac{\mu_2 \omega_2 \alpha_{123}}{\mu_1 \omega_1 \alpha_{231}}, \tag{3.170}$$

where a_2 may be isolated by means of further manipulations so that we get

$$a_2{}^2 = \frac{\mu_1 \omega_1 \omega_3}{2\alpha_{123}\alpha_{321}} \left\{ \sigma_3 \frac{\sigma + (\sigma_3/2)}{\mu_1 + \mu_2} - 2\mu_3 \right.$$
$$\pm \left[\frac{f_{33}{}^2}{\omega_3{}^2} \left(1 + \frac{(\sigma + (\sigma_3/2))^2}{(\mu_1 + \mu_2)^2} \right) \right.$$
$$\left. \left. - \left(\sigma_3 + \frac{2\mu_3(\sigma + (\sigma_3/2))}{\mu_1 + \mu_2} \right)^2 \right]^{1/2} \right\}. \tag{3.171}$$

Also we have

$$a_3 = \left(\frac{\mu_1 \mu_2 \omega_1 \omega_2}{\alpha_{123}\alpha_{321}} \right)^{1/2} \left[1 + \frac{(\sigma + (\sigma_3/2))^2}{(\mu_1 + \mu_2)^2} \right]^{1/2}. \qquad (3.172)$$

Stable solutions are defined by the real roots of equation (3.171) and it is possible, therefore, to show that one real (therefore stable) solution exists for a_2. The conditions for this are found to be

$$f_{33(\text{crit})} > \omega_3(\sigma_3{}^2 + 4\mu_3{}^2)^{1/2}, \qquad (3.173)$$

$$\sigma_3(\sigma + (\sigma_3/2)) < 2\mu_3(\mu_1 + \mu_2). \qquad (3.174)$$

Thus Fig. 3.13(a) shows a plot of response a_n against excitation f_{33} for solutions (3.170)–(3.172) given inequalities (3.173) and (3.174). The most important phenomenon is that of saturation of the directly excited mode a_3. In other words the response a_3 is independent of excitational amplitude f_{33} as long as this is above a critical value. This critical value is $f_{33(\text{crit})}$. Conversely, the indirectly excited modes (indirectly because they are excited through the internal resonance (3.143)), increase in response magnitude with increasing excitation amplitude. None of the modes respond for $f_{33} < f_{33(\text{crit})}$. This saturation phenomenon is a typical feature of systems with quadratic non-linearities, including autoparametric systems (refer to Chapter 4).

There is one further condition for which solutions are possible. This is

$$\sigma_3 \left(\sigma + \frac{\sigma_3}{2} \right) \geq 2\mu_3(\mu_1 + \mu_2). \qquad (3.175)$$

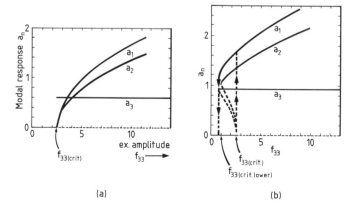

(a)   (b)

Figure 3.13 Theoretical responses for a multi-degree of freedom system with quadratic non-linearities undergoing parametric excitation: (a) note the saturation of the directly excited mode a_3 (for $f_{33} > f_{33(\text{crit})}$); (b) jump phenomena and unstable solutions arise for the following, $f_{33(\text{crit.lower})} \leq f_{33} \leq f_{33(\text{crit})}$ (after Nayfeh, 1983).

There is a lower critical excitational level

$$f_{33(\text{crit.lower})} = \frac{\omega_3 |\sigma_3(\mu_1 + \mu_2) + 2\mu_3(\sigma + (\sigma_3/3))|}{[(\mu_1 + \mu_2)^2 + (\sigma + (\sigma_3/3))^2]^{1/2}} . \qquad (3.176)$$

If $f_{33(\text{crit.lower})} \lesssim f_{33} \lesssim f_{33(\text{crit})}$ then equation (3.171) has two real solutions. For f_{33} below the lower limit there are no real solutions (other than the trivial ones – which are stable but of little interest). Values of f_{33} above the upper limit generate a single-valued stable solution for a_2 and hence a_1. Mode a_3 is seen to saturate for $f_{33} > f_{33(\text{crit.lower})}$. The case of $f_{33(\text{crit.lower})} \lesssim f_{33} \lesssim f_{33(\text{crit})}$ exhibits two solutions for a_2, the upper one is stable and the lower one is unstable. Therefore the same applies for mode a_1 (equation (3.170)). These cases are given in Fig. 3.13(b). Some interesting additional qualitative features in the form of nonlinear jump phenomena are evident here; the indirectly excited modes jumping up at the upper f_{33} limit and down at the lower limit. This could be demonstrated by variation of f_{33} within that general region. Nayfeh points out that these qualitative effects of saturation and jumping are general and are not data-specific.

Allied work on quadratic terms in nonlinear parametric equations is covered in the following; Bux and Roberts (1986), Cartmell and Roberts (1988), Nayfeh and Zavodney (1986), HaQuang and Mook (1987), Ashworth and Barr (1987). With the exception of the Nayfeh and Zavodney paper and the HaQuang and Mook paper the works quoted above are primarily concerned with auto-parametric problems in which the system as a whole comprises a forced part and a parametrically excited part, so that the parametric excitation is derived from the response of the forced part. Clearly the structural configuration will have a central role in the definition of the particular internal (i.e. parametric) resonance(s) that is/are in force. Autoparametric problems are described in detail in Chapter 4. It is particularly in Chapter 4 that we attempt to link some physical system configurations with specific equations by modelling certain systems.

(b) Parametric instabilities in a nonlinear string and fluid problem (an application of quadratic damping)

An interesting practical example of the general case of nonlinear damping in a parametric system (Section 3.2.1) is to be presented here. The original work on which this is based was published by Hsu (1975a) and concerns the lateral vibration of a long hanging string in a fluid, when subjected to longitudinal periodic excitational motion. This type of problem has been successfully used to model offshore drill string vibration problems and is therefore of great practical importance. A schematic presentation of the hanging string with moving support is

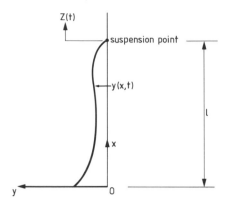

Figure 3.14 Schematic of longitudinally excited drill string (after Hsu, 1975b).

given in Fig. 3.14. Hsu points out that the string is assumed to be inextensible, and of uniform cross-section, and homogeneous material composition along its length. The string is submerged along its entire length in a fluid medium and the fluid resistance per unit length is given as

$$R = \tfrac{1}{2}\rho_f d_0 C_D \frac{\partial y}{\partial t} \left| \frac{\partial y}{\partial t} \right| \tag{3.177}$$

C_D is the drag coefficient (dependent on the fluid and the string dimensions), d_0 is effectively the string diameter, and y is the transverse displacement. It is assumed that the suspension point is excited vertically by means of a function $z(t)$

$$z(t) = z_0 \cos\left(2\pi \frac{1}{\tau_z}\right)t, \tag{3.178}$$

where $(1/\tau_z) = f$, given that τ_z is the heaving motion period. This is really just another way of saying

$$z(t) = z_0 \cos \Omega t, \tag{3.179}$$

where Ω is the frequency of excitation ($\Omega = 2\pi f$). The inertia term of the governing equation will contain the weight of the string, w_0, and also possibly the weight of other system elements, w_2 (such as internal fluid for problems of parametrically excited long flexible pipes containing a fluid for example).

Thus we can write this term as follows

$$\frac{w_0 + w_2}{g} \frac{\partial^2 y}{\partial t^2} = \text{inertia term.} \tag{3.180}$$

For a stationary suspension point the tension in the string is w_1x (w_1 being the weight per unit length of the string immersed in fluid), with x being the displacement from the lower end. For the case where the suspension point is not stationary Hsu defines the total (or net) tension in the string as

$$T_{net} = \left(w_1 + \frac{w_0}{g}\ddot{z}\right)x \qquad (3.181)$$

with w_0/g being the mass per unit length of the string which is accelerated longitudinally.

We now apply a version of equation (2.214) (refer to Section 2.4.2(a)), except that the damping term, as defined by equation (3.177), is added in to the right-hand side of our version of equation (2.214). This gives,

$$\frac{w_0 + w_2}{g}\frac{\partial^2 y}{\partial t^2} + \tfrac{1}{2}\rho_f d_0 C_D \frac{dy}{dt}\left|\frac{dy}{dt}\right| = \frac{\partial}{\partial x}\left(w_1 + \frac{w_0}{g}\ddot{z}\right)x\frac{\partial y}{\partial x}. \qquad (3.182)$$

Equation (3.182) in a nonlinear partial differential equation with a clearly defined periodic coefficient. Hsu treats this equation in several stages beginning with the linearized problem with no parametric excitation ($z(t) = 0$). By employing the' standard technique of separation of variables such that

$$y(x,\ t) = q(t)X(x) \qquad (3.183)$$

it can easily be shown that these two ordinary differential equations emerge

$$xX'' + X' + k^2X = 0, \qquad (3.184)$$

$$\left(1 + \frac{w_2}{w_0}\right)\ddot{q} + k^2\left(\frac{w_1}{w_0} + \ddot{z}\right)q = 0, \qquad (3.185)$$

where the dots and primes denote temporal and spatial derivatives respectively, and k^2 is given by

$$k^2 = -(1/X)[X' + xX'']. \qquad (3.186)$$

Solutions to equation (3.184) and (3.185) may be obtained and it is ultimately possible to deduce the natural frequency

$$\omega_n = \tfrac{1}{2}j_{0,n}\left[\frac{g(w_1/w_0)}{(1 + w_2/w_0)l}\right]^{1/2}. \qquad (3.187)$$

Note that l is the string length and $j_{0,n}$ is the nth zero of the order zero Bessel function J_0. For an explanation of the use of Bessel functions refer to McLachlan (1955). As Hsu points out there will be an eigenvector associated with the solution to equation (3.184) and each of the natural frequencies, ω_n.

Taking the Nth mode as the one that is excited into a parametric instability, so that equation (3.183) is restated,

$$y(x, t) = q_N(t)X(x). \tag{3.188}$$

Equations (3.184) and (3.185) can therefore be used to obtain an equation of motion in $q_N(\bar{\tau})$, where $\bar{\tau}$ is a dimensionless time variable

$$\bar{\tau} = (2\pi/\tau_z)t. \tag{3.189}$$

This equation of motion is

$$\frac{d^2 q_N}{d\bar{\tau}^2} + (\delta_N + \varepsilon_N \cos \bar{\tau})q_N + c_{2,N} \frac{dq_N}{d\bar{\tau}} \left| \frac{dq_N}{d\bar{\tau}} \right| = 0, \tag{3.190}$$

where

$$\delta_N = \left(\frac{T_z \omega_N}{2\pi} \right)^2 \tag{3.191}$$

and

$$\varepsilon_N = -\omega_N{}^2 z_0 \frac{w_0}{g w_1}. \tag{3.192}$$

The nonlinear damping coefficient $c_{2,N}$ is given by

$$c_{2,N} = \left[\frac{\rho_f d_0 C_D g}{2 w_0 (1 + w_2/w_0)} \right] \frac{B_{3,N}}{B_{2,N}}. \tag{3.193}$$

Note that the analysis has introduced Bessel function dependent quantities $B_{2,N}$ and $B_{3,N}$ for which numerical values have been derived for different N (Hsu, 1975a). For the fundamental ($N = 1$), these are $B_{2,1} = 0.2695$ and $B_{3,1} = 0.1949$.

Equation (3.190) is a specific case of the general problem typified by equation (3.130) in which $c_1 = \beta = 0$. Thus $c_{2,N}$ of equation (3.190) is analogous to the $c_\alpha(\alpha = 2)$ of equation (3.130), and δ_N of equation (3.190) is analogous to the δ of equation (3.130), as is the ε_N to the ε of each respective equation. Thus a form of equation (3.135) would represent the resultant response behaviour of this system

$$a^*{}_{N=1} = \frac{3\pi}{4 c_{2,N}} \sqrt{\varepsilon_N{}^2 - 4\gamma_N{}^2}; \quad (N = 1), \tag{3.194}$$

where $\gamma_N = \delta_N - (N/2)^2$.

The qualitative results would be exactly as in case (1) of Fig. 3.11.

3.3 USEFUL REFERENCES

This chapter has outlined work of the author and others, however, there is a great deal of neglected material which in one form or another could have been regarded as suitable for inclusion in this chapter. Unfortunately space constraints in an introductory text such as this preclude detailed references to much of this valuable work. The interested reader is strongly recommended to investigate the following chosen references (it should be noted that this list is by no means exhaustive): Bolotin's text, Chapter 4 entitled 'Free and forced vibrations of a non-linear system' (Bolotin, 1964); Text on rotor dynamics, Chapter 7 entitled 'Resonance vibrations of a rotor with non-linear factors taken into account' (Tondl, 1965); Paper entitled 'Parametric excitation of a non-linear system' – generalized system of the form

$$\ddot{x} + x + h_1 x^m + h_2 x^{n-1} - h_3 x^{n-1} \cos \lambda t = 0 \qquad (m, n \text{ integers})$$

(Tso and Caughey, 1965); Paper entitled 'Parametric instability of a cantilevered column with end mass' (Handoo and Sundararajan, 1971); Paper entitled 'Resonance classification in a cubic system' – contains forced and parametric terms and also cubic 'stiffness' terms (Ness, 1971); Paper entitled 'Limit cycle oscillations of parametrically excited second-order nonlinear systems; (Hsu, 1975); Text on resonance oscillations, Chapter 1, giving a general overview of non-linearities and their sources (Evan-Iwanowski, 1976); Review paper entitled 'Parametric vibration Part II: Mechanics of nonlinear problems' (Ibrahim and Barr, 1978); Text on nonlinear oscillations, various chapters (Nayfeh and Mook, 1979); Paper entitled 'A global analysis of a nonlinear systems under parametric excitation' (Guttalu and Hsu, 1984); Paper entitled 'The response of two degree systems with quadratic non-linearities to a combination parametric resonance' – probably more appropriate to Chapter 4 of this book since it deals with problems that are usually classed as autoparametric (Zavodney and Nayfeh, 1988); Text on nonlinear oscillations, Chapter 5, generalized equation containing linear, cubic ($C\dot{x}^2\dot{x}$) and quintic ($D\dot{x}^4\dot{x}$) damping terms, cubic (Ex^3) and quintic (Fx^5) stiffness non linearities and linear forced excitation, and linear and nonlinear parametric excitation terms (the latter varying with x^2) (Schmidt and Tondl, 1986); Paper entitled 'The resonance of structures with quadratic inertial non-linearities under direct and parametric harmonic excitation' (Ashworth and Barr, 1987); Paper entitled 'Non-linear structural vibrations under combined parametric and external excitations' – dealing with a multi-degree of freedom system (HaQuang and Mook, 1987); Paper entitled 'A nonlinear analysis of the interaction between parametric and external excitations' – also treats quadratic and cubic non-linearities within the same system (HaQuang and Mook, 1987); Paper entitled 'Parametric excitation of two internally resonant oscillators' (Nayfeh, 1987).

4

Nonlinear vibrations in autoparametric systems

INTRODUCTION

Autoparametric systems are distinct from parametric systems because of the internal coupling that exists in the form of nonlinear product terms involving at least two modes. In the simplest case an autoparametric system can be thought of in two parts, the first being an externally excited forced oscillator of some description, and the second comprising an oscillator parametrically excited by the response of the forced element. If we consider each part of the system to possess a single degree of freedom and if we excite the forced (or primary) part at, or around, its linearly resonant point, so that $\Omega \simeq \omega_1$, then the parametric (or secondary) part will respond (in principal parametric resonance) if we find that its excitation frequency (which will be at ω_1) is close to twice its own particular frequency, say ω_2. Therefore $\omega_1 \simeq 2\omega_2$, is identified as a critical condition, and for a structural configuration coupled in such a way that this resonance condition holds, there may be an energy flow from the source of excitation right through the whole system culminating in responses in both modes – at ω_1 and ω_2.

Furthermore, because the primary system excites the secondary system into motion there will be an acting back of the secondary system on the primary system as the secondary system responds. This will of course modify the primary response (i.e. the secondary system's own source of energy) and in so doing a steady state situation is likely to follow. We use the word likely because under certain circumstances non-steady chaotic responses can be generated. A general note on chaos in vibration problems is made in Chapter 5. For now we will continue with an investigation into steady-state behaviour in autoparametric problems of various sorts.

As has been stated above the fundamental requirement of an autoparametric system is coupling between modes (either two or more) in such a way that response relationships can be made to apply between the natural (or modal) frequencies and also the frequency of external excitation. Such a system might well be modelled in a general sense by the following equations of motion

$$\ddot{x} + 2\xi_1\omega_1\dot{x} + \omega_1{}^2x - \varepsilon\mu(\dot{y}^2 + y\ddot{y}) = P_0\cos\Omega t, \tag{4.1}$$

$$\ddot{y} + 2\xi_2\omega_2\dot{y} + \omega_2{}^2y - \varepsilon\ddot{x}y = 0, \tag{4.2}$$

where the coupling between the two equations is apparent in the terms $\varepsilon\mu(\dot{y}^2 + y\ddot{y})$ and $\varepsilon\ddot{x}y$. Equation (4.1) can be reduced to a linear forced oscillator (with one degree of freedom) if the nonlinear term $\varepsilon\mu(\dot{y}^2 + y\ddot{y})$ disappears. The other equation of motion, (4.2), is clearly a parametric type equation where $\ddot{x}$ acts as a coefficient of the coordinate y. The link between these two equations in the form of coordinate coupling is established if the following apply

$$\Omega \simeq \omega_1 \; ; \tag{4.3}$$

$$\omega_1 \simeq 2\omega_2. \tag{4.4}$$

Interaction is negligible unless resonance conditions (4.3) and (4.4) are operating.

Further resonances involving more frequencies and more complicated interrelationships than the above are possible in larger and more intricate systems. Certain cases will be examined subsequently, and some interesting phenomena will be highlighted such as steady-state modal responses at frequencies remote from the excitation frequency (i.e. as in the case of resonances (4.3) and (4.4) where $\Omega \neq \omega_2$ for example), mode saturation and modal absorption, and also response discontinuities and jumps.

4.1 TWO MODE INTERACTION IN A COUPLED BEAM SYSTEM

A very well-known type of autoparametric problem to which equations (4.1) and (4.2) apply comprises a pair of connected beams as illustrated in Fig. 4.1. The structure could be regarded as modelling (perhaps rather loosely!) an airframe fuselage and tailplane interaction where the fuselage is the directly excited primary system and the tailplane is the coupled secondary system. For the sake of simplicity and clarity we contain ourselves to a lumped two degree of freedom arrangement with each beam vibrating in its fundamental mode. A generalized analytical treatment for just this case has been given (Roberts and Cartmell, 1984). We can discount the $\varepsilon\gamma x^2$ term included in the original work

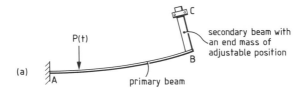

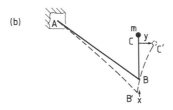

Figure 4.1 Coupled beam interaction problem showing: (a) primary response resulting from an imposed external excitation $P(t)$; and (b) schematic showing both primary and secondary displacements (both in fundamental bending modes).

(because its effects on the system are known to be negligible) and consider, therefore, equations (4.1) and (4.2) as being adequately descriptive. The damping and excitation terms are to first-order ε; $\xi_{1,2} = \varepsilon\zeta_{1,2}$ and $P_0 = \varepsilon p$. As is almost always the case with nonlinear differential equations simple analysis is not possible, and so an approximate analysis such as multiple scales needs to be used. By taking the usual series forms for the coordinates and their first and second derivatives, and polar notation for the complex amplitudes, a set of four slow time first-order equations may be constructed by the removal of secular terms and the utilization of the resonance conditions. The resonance conditions are, of course, derived from equations (4.3) and (4.4) introducing the detuning parameters $\varepsilon\rho_1$ and $\varepsilon\rho_2$, thus,

$$\Omega = \omega_1 + \varepsilon\rho_1; \tag{4.5}$$

$$\omega_1 = 2\omega_2 + 2\varepsilon\rho_2. \tag{4.6}$$

The steady-state amplitude responses (based on the slowly varying, polar amplitudes, a' and b' being static; $a' = b' = 0$) may be derived, and give two cases; firstly no interaction (i.e. $y = b = 0$), in which situation equation (4.1) drops the nonlinear coupling term and becomes the equation for a forced linear oscillator, with a solution as follows

$$a = \frac{P_0}{2\omega_1} [(\Omega - \omega_1)^2 + \xi_1{}^2\omega_1{}^2]^{-1/2}, \tag{4.7}$$

$$b = 0. \qquad (4.8)$$

Equations (4.7) and (4.8) express how the system behaves when there is no interaction between the two modes.

The other possibility is that interaction does occur and that equation (4.8) does not therefore apply. In this case the primary response becomes

$$a = \frac{4\omega_2^2}{\varepsilon\omega_1^2} \left[\left(\frac{\Omega}{2\omega_2} - 1 \right)^2 + \xi_2^2 \right]^{1/2} \qquad (4.9)$$

and the secondary response is deducible from the following quadratic,

$$b^4 + Qb^2 + R = 0, \qquad (4.10)$$

where Q and R are defined by

$$Q = \frac{16}{\varepsilon^2 \mu \omega_1 \omega_2} \left[\xi_1 \xi_2 \omega_1 \omega_2 - (\Omega - \omega_1) \left(\frac{\Omega}{2} - \omega_2 \right) \right], \qquad (4.11)$$

$$R = \frac{64}{\varepsilon^4 \mu^2 \omega_1^2 \omega_2^2} \left[((\Omega - \omega_1)^2 + \xi_1^2 \omega_1^2) \right.$$

$$\left. \times \left(\left(\frac{\Omega}{2} - \omega_2 \right)^2 + \xi_2^2 \omega_2^2 \right) \right] - \frac{P_0^2}{\varepsilon^2 \mu^2 \omega_2^4}. \qquad (4.12)$$

The quadratic equation (4.10) in b^2 will potentially give zero, one or two real roots which represent steady vibrations of the coupled beam BC at frequency ω_2.

The interactive region is bounded by ABCD on the curves of Fig. 4.2(a) and 4.2(b). Taking Fig. 4.2(a) first we find that AGEFD and BGEFC both represent stable solution paths, whereas GHF is unstable. The linear curve is, of course, stable at all times outside AD, and the nonlinear curve is unstable outside jumps AB and CD. For the secondary responses of Fig. 4.2(b) the zero solution (i.e. $b = 0$) is stable along HAG and FDI but becomes unstable between GF. The non-zero solution ($b \neq 0$) is stable within BEC but unstable outside the verticals AB and CD.

So taken together equations (4.7), (4.9) and (4.10) model the linear behaviour and the nonlinear interactive behaviour that can occur when resonance conditions (4.5) and (4.6) are operative. It is an interesting characteristic of the autoparametric interaction case (equations (4.9), (4.10)) that the excitation function P_0 does not appear in the expression for primary response (equation (4.9)), whereas it is evident (as one would expect) in the linear solution given by equation (4.7). The only point at which it does appear in the nonlinear interaction case is the final term of equation (4.10), defining the secondary response (refer also

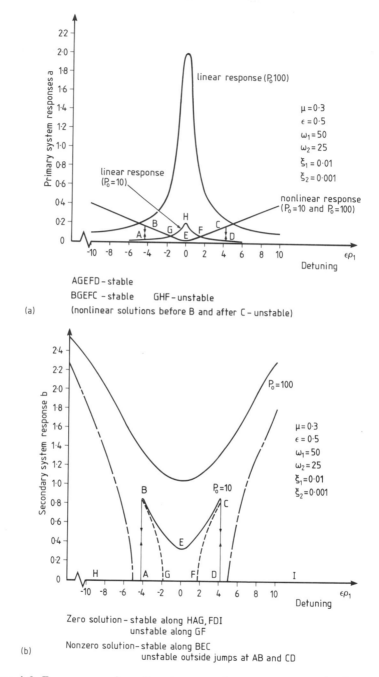

Figure 4.2 Responses for the two mode autoparametric interaction problem: (a) theoretical primary system responses; (b) theoretical secondary system responses.

to equation (4.12)). This suggests that the primary response is not governed by the excitational function as long as interaction between the modes is occurring. This is an intriguing situation since it is of course the primary system itself that is directly under the influence of the excitation P_0. The physical outcome of this can be a suppression of the directly excited primary mode under certain conditions. This is called the saturation effect, and is further examined later on in this chapter (refer to Section 4.3.1).

Typical solutions, as functions of frequency (detuning $\varepsilon\rho_1$) are given in Figs 4.2 and 4.3. Taking the primary system responses of Fig. 4.2(a), where there is an exact $2:1$ relationship between ω_1 and ω_2, we see that the two linear responses chosen as examples ($P_0 = 10$ and $P_0 = 100$) are replaced by the shallow vee-shaped curve when nonlinear autoparametric interaction is allowed to occur. The externally resonant point at $\Omega = \omega_1$ (or $\varepsilon\rho_1 = 0$) shows a maximum linear response and a corresponding minimum non-linear secondary response. It can be seen that irrespective of the precise value of P_0 large amounts of energy are extracted from the primary mode (particularly at external resonance) when autoparametric interaction is underway.

Figure 4.2(b) illustrates the secondary system responses, again for the two chosen excitation levels.

The upper branches of the curves (solid line) are stable solutions whereas the lower branches (chain and dotted lines) are unstable, and in practice a system would keep to the upper solutions at these points. By steadily increasing the frequency from a remote (low) point, say $\varepsilon\rho_1 = -10$ in the case of $P_0 = 10$, we would find that the system follows the linear response curve of Fig. 4.2(a) and that there is initially no interaction and therefore no secondary response. However at $\varepsilon\rho_1 \simeq -4.3$ both parts of the system suddenly show large qualitative and quantitative changes in their respective responses in the form of jumps from A to B. As the frequency sweep is continued each response travels along the non-linear curves from B, through and then past external resonance at E until point C is reached. Continuing the frequency sweep still further would result in a sudden return to linear response behaviour as the curves jump downwards from C to D. The jumps AB and CD are symmetrical about $\varepsilon\rho_1 = 0$ and of the same magnitude because this particular example is for a perfectly tuned system (i.e. the internal resonance is exact at $\omega_1 = 2\omega_2$). A larger excitation level (say $P_0 = 100$) would extend the frequency region over which the nonlinear characteristics occur and would increase the magnitudes of the responses. Note however that the nonlinear primary response curve is independent of P_0, and that an increase in P_0 to $P_0 = 100$ would simply promote jumps at points further away from $\varepsilon\rho_1 = 0$ than the lower excitational level of $P_0 = 10$.

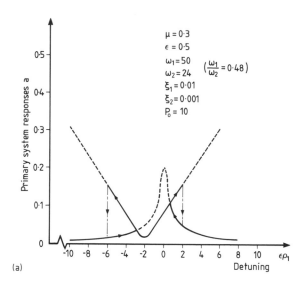

(a)

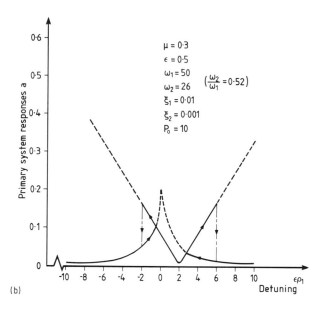

(b)

Figure 4.3 Theoretical primary system responses for the internally detuned system: (a) $((\omega_2/\omega_1) = 0.48)$; (b) $((\omega_2/\omega_1) = 0.52)$; (c) theoretical secondary system responses for two cases of internal detuning $((\omega_2/\omega_1) = 0.48$ and $0.52)$.

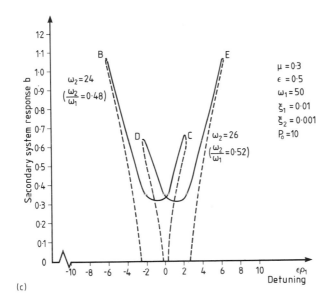

(c)

Reducing the system damping levels will tend to increase the non-linear region both in frequency span and magnitude of responses.

The system will also operate effectively in the nonlinear domain for small amounts of internal detuning; i.e. when $\omega_2/\omega_1 \simeq 0.5$. Two cases are outlined here and results are given in Figs 4.3(a), (b) and (c). Figure 4.3(a) illustrates the case of $\omega_2/\omega_1 = 0.48$ and shows how the nonlinear primary response curve offsets in such a way that it is symmetrical about $\varepsilon\rho_1 < 0$ (around $\varepsilon\rho_1 \simeq -2$). The qualitative characteristics of the response are in all other ways unchanged, however the overall effect is that the linear response is not now symmetrically absorbed by the nonlinear response. The secondary system response curve for $\omega_2/\omega_1 = 0.48$ is shown in Fig. 4.3(c). The symmetry of the curve is no longer apparent and the extremal responses at the jump points (these are the points of vertical tangency at B and C on Fig. 4.2(b) and B, C, D and E on Fig. 4.3(c)) reach a higher level at B than C (Fig. 4.3(c)). Behaviour of the secondary system outside the autoparametric region is identical for all cases (irrespective of the degree of internal detuning) and is such that no vibratory motion occurs at all (zero responses of b). The other case of internal detuning ($\omega_2/\omega_1 = 0.52$) offsets to the other (+ve) side of $\varepsilon\rho_1 = 0$; refer to Figs 4.3(b) and 4.3(c). Qualitatively the responses are identical to the $\omega_2/\omega_1 = 0.48$ case, except with a reversed asymmetry.

Higher levels of internal detuning will force the system into more radical response asymmetries until eventually the two parts will decouple, and the overall system behaviour will once again be linear over the whole of the active excitation frequency range (i.e. the range around $\varepsilon\rho_1 = 0$ in which the system is found to respond). Thus there will then be a purely linear response of the primary system and a zero response of the secondary system. Nonlinear behaviour could be re-initiated for extreme cases of internal detuning if the system excitation level was increased sufficiently. Another way of returning to nonlinear response behaviour for the system under conditions of extreme internal detuning would be to drastically reduce the values of ξ_1 and ξ_2. In most practical situations such reduced damping levels would be unlikely to be deliberately accommodated in design work!

4.2 APPLICATION TO VIBRATION ABSORPTION

Use has already been made in the previous section of the word absorption in the context of energy transfer from the primary system mode to the secondary mode(s). In this section we attempt to show how direct exploitation of this phenomenon can enable us to analyse, design and construct mechanical vibration absorbers in which autoparametric interaction provides the energy transfer mechanism. The first publication of such work known to the author, was that due to Haxton and Barr (1972) in which they described a two degree of freedom system comprising a linear spring–mass system (subjected to external forcing) and an attached absorber beam which is excited parametrically by the spring–mass system. Nonlinear absorber motion terms act back on the spring–mass primary system so that absorption of its motion may be initiated by careful choice of tuning parameters. The schematic of the system is given in Fig. 4.4 and the experimental rig is shown in Fig. 4.5.

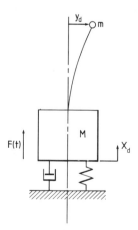

Figure 4.4 Autoparametric absorber schematic (after Haxton and Barr, 1972).

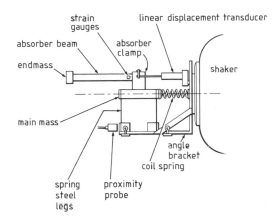

Figure 4.5 Autoparametric absorber experimental apparatus (after Haxton and Barr, 1972).

The equations of motion are similar in form to the general pair given in the introduction (equations (4.1) and (4.2)) and are derived by means of the method of Lagrange. The original authors chose to non-dimensionalize these and to treat them by Struble's method of asymptotic approximations (Struble, 1962) in order to get non-trivial non-linear responses.

Taking the schematic of Fig. 4.4 we can write down the system kinetic energy

$$T = \tfrac{1}{2}M\dot{X}_\mathrm{d}^2 + \tfrac{1}{2}m(\dot{u}_\mathrm{d} + \dot{X}_\mathrm{d})^2 + \tfrac{1}{2}m\dot{y}_\mathrm{d}^2. \tag{4.13}$$

Timoshenko's relationship between the general axial contraction u and the general lateral displacement y for a cantilevered flat spring, or alternatively a flat beam of length l, can be introduced, such that,

$$u = \tfrac{1}{2}\int_0^l y'^2\,\mathrm{d}x. \tag{4.14}$$

Equation (4.14) may be more usefully stated if we take the normalized deflection form function for a cantilever

$$\phi(x) = \frac{x^2}{2l^2}(3 - x); \qquad (\phi(l) = 1) \tag{4.15}$$

and incorporate this into equation (4.14) to give

$$u_\mathrm{d} = \tfrac{1}{2}y_\mathrm{d}^2\int_0^l \phi'(x)^2\,\mathrm{d}x \tag{4.16}$$

finally leading to

$$u_\mathrm{d} = \frac{6}{5l}y_\mathrm{d}^2 \tag{4.17}$$

(the suffix d being introduced since these deflections occur at $x = l$, refer to Fig. 4.4).

The contraction velocity, is therefore,

$$\dot{u}_\mathrm{d} = \frac{6}{5l}\, y_\mathrm{d}\dot{y}_\mathrm{d}. \tag{4.18}$$

So the system kinetic energy is now

$$T = \tfrac{1}{2}M\dot{X}_\mathrm{d}^2 + \tfrac{1}{2}m\left(\frac{6}{5l}\, y_\mathrm{d}\dot{y}_\mathrm{d} + \dot{X}_\mathrm{d}\right)^2 + \tfrac{1}{2}m\dot{y}_\mathrm{d}^2. \tag{4.19}$$

It is fairly clear that expansion of the second, bracketed, term is going to generate intercoordinate coupling, such as will eventually give rise to typical autoparametric quadratic non-linearities, once the Lagrangian derivation has been completed.

The strain energy terms are straightforward

$$U = \tfrac{1}{2}KX_\mathrm{d}^2 + \tfrac{1}{2}\lambda y_\mathrm{d}^2. \tag{4.20}$$

Haxton and Barr point out that it is not necessary to restrict the physical design of the absorber (or secondary) element to a cantilevered flat spring, and alternative designs are possible. Some current work is underway in which an adjustable length vertical pendulum is used instead. The pendulum is restrained to the neutral axial position by means of springs. Schematic and experimental configurations are given in Figs 4.6 and 4.7. For this design the system energies are of the forms

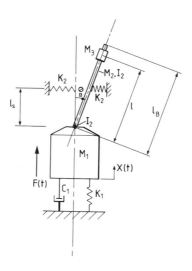

Figure 4.6 Schematic of an autoparametric absorber using a pendulum.

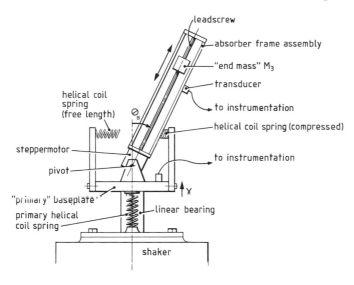

Figure 4.7 Experimental pendulum absorber.

$$T = \tfrac{1}{2}M_1\dot{X}^2 + \tfrac{1}{2}I_1\dot{\theta}_B{}^2 + \tfrac{1}{2}I_2\dot{\theta}_B{}^2 + \tfrac{1}{2}M_2[(\dot{X} - \tfrac{1}{2}l_B\dot{\theta}_B \sin \theta_B)^2$$
$$+ (\tfrac{1}{2}l_B\dot{\theta}_B \cos \theta_B)^2] + \tfrac{1}{2}M_3[(\dot{X} - l\dot{\theta}_B \sin \theta_B)^2 + (l\dot{\theta}_B \cos \theta_B)^2], \quad (4.21)$$
$$U = \tfrac{1}{2}K_1 X^2 + \tfrac{1}{2}K_2 l_s{}^2 \tan^2 \theta_B. \quad (4.22)$$

Applying the method of Lagrange to equations (4.21) and (4.22) requires the use of the small angle approximations, $\cos \theta_B \simeq 1$, $\sin \theta_B \simeq \theta_B$; $\tan \theta_B \simeq \theta_B$ for small θ_B in order to keep the algebra within manageable limits. The equations of motion which result are identical in form to equations (4.1) and (4.2)

$$\ddot{X} + 2\xi_1\omega_1\dot{X} + \omega_1{}^2 X - \varepsilon h(\theta_B\ddot{\theta}_B + \dot{\theta}_B{}^2) = \varepsilon f \cos \Omega t, \quad (4.23)$$

$$\ddot{\theta}_B + 2\xi_2\omega_2\dot{\theta}_B + \omega_2{}^2\theta_B - \varepsilon g\ddot{X}\theta_R = 0, \quad (4.24)$$

where the various coefficients and system constants are

$$\xi_1 = \frac{c_1}{2\omega_1(M_1 + M_2 + M_3)}; \qquad \omega_1{}^2 = \frac{K_1}{M_1 + M_2 + M_3}, \quad (4.25)$$
$$(4.26)$$

$$\varepsilon = \frac{l_B}{2}(M_2 + lM_3); \qquad h = \frac{1}{M_1 + M_2 + M_3}, \quad (4.27)$$
$$(4.28)$$

$$\xi_2 = \frac{C_2}{2\omega_2[I_1 + I_2 + ((l_B{}^2/4)M_2 + l^2M_3)]}, \quad (4.29)$$

$$\omega_2^2 = \frac{K_2 l_s^2}{[I_1 + I_2 + ((l_B^2/4)M_2 + l^2 M_3)]},$$ (4.30)

$$g = \frac{1}{[I_1 + I_2 + ((l_B^2/4)M_2 + l^2 M_3)]},$$ (4.31)

$$f = \frac{F_0}{[(M_1 + M_2 + M_3)((l_B/2)M_2 + lM_3)]}.$$ (4.32)

The solutions to equations (4.23) and (4.24) will be identical to those of equations (4.1) and (4.2), however εh will replace $\varepsilon \mu$, and the term εg will appear in the denominator of the non-linear primary response, thus,

$$a = \frac{4\omega_2^2}{\varepsilon g \omega_1^2} \left[\left(\frac{\Omega}{2\omega_2} - 1 \right)^2 + \xi_2^2 \right]^{1/2}$$ (4.33)

This should be compared with equation (4.9). ε is a perturbation parameter and it is also an explicit system parameter defined in equation (4.27). Damping is treated to first-order ε again so that $\xi_{1,2} = \varepsilon \zeta_{1,2}$.

The nonlinear secondary responses (or absorber responses) are represented by equation (4.10) but with slightly revised coefficients so that here we have the following

$$Q = \frac{16}{\varepsilon h \varepsilon g \omega_1 \omega_2} \left[\omega_1 \omega_2 \xi_1 \xi_2 - (\Omega - \omega_1) \left(\frac{\Omega}{2} - \omega_2 \right) \right],$$ (4.34)

$$R = \frac{64}{(\varepsilon h)^2 (\varepsilon g)^2 \omega_1^2 \omega_2^2} [(\Omega - \omega_1)^2 + \xi_1^2 \omega_1^2] \left[\left(\frac{\Omega}{2} - \omega_2 \right)^2 + \xi_2^2 \omega_2^2 \right]$$
$$- \frac{F_0^2}{\varepsilon^2 \omega_2^4}.$$ (4.35)

Figures (4.2) and (4.3) provide master plots for the (steady state) qualitative behaviour of the pendulum type autoparametric absorber. Further work on this design of absorber is to be published by the author in due course. For now it can be concluded that while the absorption of the autoparametric type of device is powerful it does not, at first, seem to be a more efficient performer than the more conventional tuned and damped device in terms of its potential to transfer vibrational energy. Where it does appear to have an advantage is in its ability to lock on to nonlinear (i.e. absorbing) responses over the range of forcing frequencies between the jumps, and also that small amounts of internal detuning (ω_2/ω_1) may be used, with advantage, to optimize absorber action in certain cases of external detuning. Dimension l is variable (Fig. 4.6) for this purpose. An interesting adaptive absorber is at least potentially realizable by exploiting these facets of the system.

4.3 MULTIMODAL AUTOPARAMETRIC INTERACTIONS

4.3.1 Three mode interaction driving an internal combination resonance

The coupled beam system of Section 4.1 (Fig. 4.1) has been studied further (Bux and Roberts, 1986) with the intention of introducing a combination resonance of the form of equation (2.74). To this end three modal coordinates have been taken; these represent motion of the primary beam in its second bending mode and motion of the secondary beam in its fundamental bending and torsion modes. The second primary (or planar) bending mode is chosen since it provides both a vertical translation and a planar rotation at the coupling point, both of which can be shown to be contributory to a driven parametric combination resonance in the secondary beam. The forced linear response of the primary beam vibrating in its second bending mode is typically as in Fig. 1.26, where the node is approximately at $0.8l$.

Taking X_2 as the non-dimensionalized coordinate for planar (primary) motion in the second bending mode (hence subscript 2) and Y_1 and Y_2 as non-planar (secondary) non-dimensionalized motion coordinates in fundamental bending and torsion respectively, the three equations of motion can be derived. We will adhere to some of the originators' notation and it should be noted that all nonlinear terms, damping terms and the external excitation term are ordered to ε. The underlying equation for the planar system is

$$[M]\ddot{\bar{q}} + [C]\dot{\bar{q}} + [K]\bar{q} = [N], \tag{4.36}$$

where $[N]$ is a vector of forces comprising interactive coupling point force terms and any externally applied forces. If the force application points are denoted by subscripts i representing locations and associated with coordinates q_i, then this vector will be of the following form, given the structural configuration of Fig. 4.1,

$$[N] = \begin{bmatrix} N_1 \\ N_2 \\ \vdots \\ N_i \\ \vdots \\ N_k \end{bmatrix}; \quad \text{where the non-zero elements are } N_1, N_2 \text{ and } N_k.$$

$$\tag{4.37}$$

N_1 is the force acting at the coupling point due to interaction between the primary beam's translation at the coupling point and the non-planar bending of the secondary beam. N_2 is the moment at the coupling point due to the combined non-planar bending and torsional motion of the secondary beam. N_k is the externally imposed excitation force at some chosen station (designated by k, where k is some discrete position along the primary beam).

A generalized coordinate system is introduced which can be identified with specific modes by means of the following transformation

$$q_i = \psi_{i1} X_1 + \psi_{i2} X_2 \tag{4.38}$$

and noting that since our concern is exclusively with the second planar mode we reduce equation (4.38) to

$$q_i = \psi_{i2} X_2, \tag{4.39}$$

where ψ_{i2} is the ith element of the second eigenvector of the linear undamped motion of the whole structure; so now we can apply the technique of Lagrange and write the basic equation of planar motion, inserting the appropriate elements of the column vector $[N]$

$$M_j \ddot{X}_j + K_j X_j = \psi_{kj} P_0 \cos \Omega t + \psi_{1j} m \int_0^l [f']^2 \, \mathrm{d}z (u_0 \ddot{u}_0 + \dot{u}_0{}^2)$$

$$+ \psi_{2j} m \int_0^l (l - z) f'' g \, \mathrm{d}z (\ddot{u}_0 \phi_0 + 2\dot{u}_0 \dot{\phi}_0 + u_0 \ddot{\phi}_0), \tag{4.40}$$

where M_j and K_j are the generalized mass and stiffness respectively, of the jth mode ($j = 2$). ψ_{kj}, ψ_{1j}, ψ_{2j} are derivable eigenvectors at stations k, 1 and 2 respectively for the linear problem (these could be obtained by recourse to a linear model using the PAFEC Finite Element package for example).

Thus, equation (4.40) is the transformed equation of planar motion for which, if $j = 2$, the coordinates define motion in the second mode of vibration. The second term on the right-hand side of this equation contains $(u_0 \ddot{u}_0 + \dot{u}_0{}^2)$ which can be shown to be negligible if we are operating well away from an internal condition of principal parametric resonance; therefore $\omega_2 \neq 2\omega_b$ (where ω_b is the fundamental, non-planar, secondary bending mode natural frequency).

Since we are in fact interested solely in a combination resonance which is well removed from $2\omega_b$ then this condition is satisfied and the term may safely be neglected. The $\omega_2 \simeq 2\omega_b$ case is discussed in detail in Section 4.1. Equation (4.40) can now be rewritten

$$\ddot{X}_2 + \omega_2{}^2 X_2 = \varepsilon [P \cos \Omega t + \alpha_{22} (\ddot{u}_0 \phi_0 + 2\dot{u}_0 \dot{\phi}_0 + u_0 \ddot{\phi}_0) - 2\zeta_2 \omega_2 \dot{X}_2],$$

$$\tag{4.41}$$

where

$$\varepsilon P = \frac{\psi_{k2} P_0}{M_2}; \qquad \varepsilon \alpha_{22} = \frac{\psi_{22} m}{M_2} \int_0^l (l - z) f'' g \, \mathrm{d}z$$

and a viscous damping term is added where ξ_2 is the damping ratio specific to vibration in the second planar bending mode (and $\xi_2 = \varepsilon \zeta_2$).

Applying Lagrange gives the two secondary system equations

$$m_0 \ddot{u}_0 + EI \int_0^l [f'']^2 \, dz u_0 = m \int_0^l [f']^2 \, dz \psi_{12} \ddot{X}_2 u_0$$

$$+ \, m \int_0^l (l - z) f'' g \, dz l \psi_{22} \ddot{X}_2 \phi_0, \qquad (4.42)$$

$$I \ddot{\phi}_0 + GJ \int_0^l [g']^2 \, dz \phi_0 = m \int_0^l (l - z) f'' g \, dz l \psi_{22} \ddot{X}_2 u_0, \quad (4.43)$$

which may be restated respectively

$$\ddot{u}_0 + \omega_b^2 u_0 = \varepsilon [\beta_{12} \ddot{X}_2 u_0 + \beta_{22} \ddot{X}_2 \phi_0 - 2 \zeta_b \omega_b \dot{u}_0], \quad (4.44)$$

$$\ddot{\phi}_0 + \omega_t^2 \phi_0 = \varepsilon [\beta_{22} \ddot{X}_2 u_0 - 2 \zeta_t \omega_t \dot{\phi}_0]. \qquad (4.45)$$

As in the case of equation (4.40) there is one potentially redundant term in equation (4.44); this is the first term of the right-hand side, $\beta_{12} X_2 u_0$, which is non-resonant in the case of a combination resonance well removed from $\omega_2 \simeq 2\omega_b$. Thus once again we invoke the condition that $\omega_2 \neq 2\omega_b$ and neglect this term. Equation (4.44) reduces to

$$\ddot{u}_0 + \omega_b^2 u_0 = \varepsilon [\beta_{22} \ddot{X}_2 \phi_0 - 2 \zeta_b \omega_b \dot{X}_2]. \qquad (4.46)$$

Equation (4.45) remains unchanged. The coupling coefficient $\varepsilon \beta_{22}$ is defined as

$$\varepsilon \beta_{22} = \frac{m \psi_{12}}{m_0} \int_0^l (l - z) f'' g \, dz.$$

It should be noted that masses m and m_0 are different, m is the actual discrete lumped mass at the end of the secondary beam and m_0 is an equivalent, or effective, mass of the secondary system when vibrating in its fundamental bending mode. In the original work by Bux and Roberts the three governing equations (4.41), (4.45) and (4.46) are part of a generalized system of four equations (this incorporates the fundamental planar bending mode, equivalent to motion in X_1 in the system described here). They are also non-dimensionalized in the original work, however this transformation is not performed here in order to simplify the ensuing explanation. The physical concepts which are brought out in the three governing equations may be described as follows; firstly we are dealing here with an externally imposed forcing of the system in the form of a discrete point-action harmonic function at station k which oscillates at Ω, this being close to the second planar bending mode frequency, therefore

$$\Omega = \omega_2 + \varepsilon \sigma. \qquad (4.47)$$

Planar response motion at ω_2 is seen to be capable of coupling with non-planar secondary bending and torsion responses at the respective modal frequencies, however such motions will not couple unless there is an appropriate resonance condition and so to this end we stipulate that the following combination resonance applies

$$\omega_2 = \omega_b + \omega_t + \varepsilon\sigma_1. \tag{4.48}$$

The interaction implied by resonance (4.48) is due to the large pitching angular acceleration at the coupling point (which in turn is due to resonance in the form of equation (4.47)) and the fact that this acceleration of the coupling point appears as a symmetrically coupled parametric load in the secondary beam non-planar bending and torsion equations (equations (4.46) and (4.45)). These terms are $\varepsilon\beta_{22}\ddot{X}_2\phi_0$ and $\varepsilon\beta_{22}\ddot{X}_2 u_0$ respectively.

Combining equations (4.47) and (4.48) shows how the external excitation can drive the internal combination resonance

$$\Omega \simeq \omega_2 \simeq \omega_b + \omega_t. \tag{4.49}$$

A multiple scales expansion of equations (4.44), (4.45) and (4.46) involving resonances (4.47) and (4.48) enables the following set of solvability equations to be written (representing the secular terms that need to be removed from the right-hand sides of the first-order perturbation equations); note that the notation has been changed somewhat from that originally used by Bux and Roberts,

$$\omega_2 a' = -\zeta_2\omega_2{}^2 a + (\alpha_{22}/4)(\omega_b + \omega_t)^2 c_1 c_2 \sin\gamma_2 + (P/2)\sin\gamma_1, \tag{4.50}$$

$$\omega_2 a\gamma_1' = \omega_2\sigma a - (\alpha_{22}/4)(\omega_b + \omega_t)^2 c_1 c_2 \cos\gamma_2 + (P/2)\cos\gamma_1, \tag{4.51}$$

$$\omega_b c_1' = -\zeta_b\omega_b{}^2 c_1 - (\beta_{22}/4)\omega_2{}^2 c_2 a \sin\gamma_2, \tag{4.52}$$

$$\omega_t c_2' = -\zeta_t\omega_t{}^2 c_2 - (\beta_{22}/4)\omega_2{}^2 c_1 a \sin\gamma_2, \tag{4.53}$$

$$c_1 c_2 \omega_b \omega_t (\gamma_1 + \gamma_2)' = \omega_b\omega_t(\sigma + \sigma_1)c_1 c_2$$
$$- (\beta_{22}/4)\omega_2{}^2(\omega_b c_1{}^2 + \omega_t c_2{}^2)a \cos\gamma_2, \tag{4.54}$$

where the zeroth-order perturbation solutions have been defined as

$$X_{20} = A\exp(i\omega_2 T_0) + \text{C.C.} \tag{4.55}$$

$$u_{10} = C_1\exp(i\omega_b T_0) + \text{C.C.} \tag{4.56}$$

$$\phi_{10} = C_2\exp(i\omega_t T_0)\ \text{C.C.} \tag{4.57}$$

and the complex amplitudes A, C_1, C_2 are expressed in their polar forms

$$A = (a/2)\exp(i\eta); \tag{4.58}$$

$$C_1 = (c_1/2) \exp(i\mu_1); \tag{4.59}$$

$$C_2 = (c_2/2) \exp(i\mu_2) \tag{4.60}$$

also the phase angles are found to be

$$\gamma_1 = \sigma T_1 - \eta; \tag{4.61}$$

$$\gamma_2 = \sigma_1 T_1 - \mu_1 - \mu_2 + \eta. \tag{4.62}$$

Equations (4.50)–(4.54) are slow time first-order differential equations where the variables a, c_1, c_2, γ_1, γ_2 are changing with respect to the slow time-scale T_1. Multiplication of each term of all five equations by ε would transform the independent variable (i.e. the time-scale) back to $T_0 (T_0 \simeq t)$. This process also enables a return to actual, definable, system constants in the form of $\varepsilon\sigma$, $\varepsilon\sigma_1$, $\varepsilon\alpha_{22}$, εP, ξ_2, ξ_b, ξ_t, $\varepsilon\beta_{22}$.

By setting the slowly moving left-hand sides of equations (4.50)–(4.54) to zero we impose conditions through which steady-state solutions may be investigated. As in the previous discussion two outcomes are possible; firstly planar motion only (or a purely linear response of the system) for which $a \neq 0$, $c_1 = c_2 = 0$. In this case equations (4.50)–(4.54), after multiplication of each term by ε, reduce to

$$a = \frac{\varepsilon P}{2\omega_2} [(\varepsilon\sigma)^2 + \xi_2^2\omega_2^2]^{-1/2}. \tag{4.63}$$

Equation (4.63) describes a forced linear response of the second planar bending mode. This is directly comparable with equation (4.7). The second case is where non-planar bending and torsion are not precluded, therefore $a \neq 0$, $c_1 \neq 0$, $c_2 \neq 0$. This is the nonlinear case.

For planar motion we now have

$$a \frac{4\omega_b\omega_t}{\varepsilon\beta_{22}\omega_2^2} \sqrt{\xi_b\xi_t} \sqrt{1 + \frac{(\varepsilon\sigma + \varepsilon\sigma_1)^2}{(\xi_b\omega_b + \xi_t\omega_t)^2}}. \tag{4.64}$$

The non-planar (secondary) bending response is given by

$$c_1^2 = \frac{2}{\varepsilon\alpha_{22}(\omega_b + \omega_t)^2} \sqrt{\frac{\xi_t\omega_t^2}{\xi_b\omega_b^2}} [+\sqrt{(\varepsilon P)^2 - K_2^2} - K_1], \tag{4.65}$$

where

$$K_1 = \left[\frac{8\sqrt{\xi_b\xi_t\omega_b^2\omega_t^2}}{\varepsilon\beta_{22}(\xi_b\omega_b + \xi_t\omega_t)\omega_2} \right] [\xi_2\omega_2(\xi_b\omega_b + \xi_t\omega_t) - \varepsilon\sigma(\varepsilon\sigma + \varepsilon\sigma_1)],$$

$$\tag{4.66}$$

$$K_2 = \left[\frac{8\sqrt{\xi_b\xi_t\omega_b^2\omega_t^2}}{\varepsilon\beta_{22}(\xi_b\omega_b + \xi_t\omega_t)\omega_2} \right] [(\xi_b\omega_b + \xi_t\omega_t)\varepsilon\sigma + \xi_2\omega_2(\varepsilon\sigma + \varepsilon\sigma_1)].$$

$$\tag{4.67}$$

And finally the secondary torsional response is defined by

$$c_2 = \sqrt{\frac{\xi_b \omega_b^2}{\xi_t \omega_t^2}}\, c_1. \tag{4.68}$$

In the non-planar case it is clear that the bending mode c_1 may have zero, or one or two real solutions that are non-zero. Therefore so must c_2 (equation (4.68)).

Solution stability is generally investigated by looking at the effect of a small perturbation added to the solutions, from which a stability determinant is constructed. Generally speaking if there is only one solution for the non-planar mode(s) then it/they will be stable, but if there are two then the upper one is likely to be stable and the lower one to be unstable.

As in the two previous cases the planar mode is independent of the excitation function showing that the saturation effect will occur for the resonances of this case too. The original authors point out that the primary response is also independent of primary system damping ξ_2; this is also in keeping with the two previous examples that have been examined.

Graphs of planar and non-planar responses based on equations (4.63) to (4.68) are given in Figs 4.8–4.10.

Two cases are cited, one for perfect tuning of the internal resonance (therefore $\varepsilon\sigma_1 = 0$ in equation (4.48)) and one for which this combination resonance is detuned ($\varepsilon\sigma_1 = 1.5$). Taking a gradually increasing frequency sweep for the perfectly tuned case; the linear primary solution, a given in Fig. 4.8, is stable from A to B, at which point it

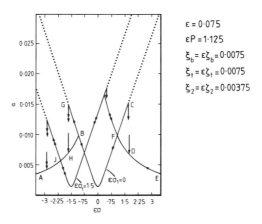

Figure 4.8 Theoretical primary response for three mode autoparametric interaction problem; two detuning cases $\varepsilon\sigma_1 = 0$ and $\varepsilon\sigma_1 = 1.5$ (after Bux and Roberts, 1986).

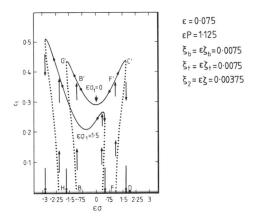

Figure 4.9 Theoretical secondary mode response (c_1) for three mode autoparametric interaction problem; two detuning cases $\varepsilon\sigma_1 = 0$ and $\varepsilon\sigma_1 = 1.5$ (after Bux and Roberts, 1986).

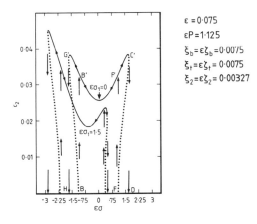

Figure 4.10 Theoretical secondary mode response (c_2) for three mode autoparametric interaction problem; two detuning cases $\varepsilon\sigma_1 = 0$ and $\varepsilon\sigma_1 = 1.5$ (after Bux and Roberts, 1986).

becomes unstable and the nonlinear response suddenly becomes stable and takes over. This remains the case until point C whereupon the nonlinear solution becomes unstable and jumps down to D, the linear solution. The planar system continues in this linear state for further increases in excitation frequency. A reversed frequency sweep has exactly the same qualitative effect except that the system travels along EFBGHA instead of ABFCDE.

To trace the behaviour of the non-planar modes c_1 and c_2 we return to the increasing frequency sweep and see that the jump from the zero solution to the non-zero solution is at BB′ and the corresponding downward jump from non-zero to zero solution is at C′D. This is the case for both the non-planar bending and torsion modes c_1, and c_2. Travelling the other way we get an upward jump at FF′ and a downward jump at G′H.

The internally detuned case ($\varepsilon\sigma_1 = 1.5$) is qualitatively identical in behaviour however the curves are no longer symmetrical about $\varepsilon\sigma = 0$.

An alternative portrayal of system behaviour is to fix the external and internal detuning levels (i.e. $\varepsilon\sigma$ and $\varepsilon\sigma_1$), so that they are constant, and then investigate the effects of altering the excitation parameter εP. Two cases are shown in Figs 4.11 to 4.13; case (1) is for $\varepsilon\sigma = 0$ and case (2) is for $\varepsilon\sigma = 1.875$ (in both cases $\varepsilon\sigma_1 = 0$). In case (1) the linear solution for the planar mode, a, is only apparent for a small region of $\varepsilon P > 0$. By the time point F is reached this solution has become unstable and the mode saturates for further increases in εP (F to G). The two non-planar modes do not response until point F is reached after which there is a smooth increase in both c_1 and c_2 which are now non-zero and stable. A reduction of εP towards zero shows identical actions in reverse, thus there is no hysteretical behaviour for the $\varepsilon\sigma = 0$ case.

Case (2) is rather more complicated and there is a rather larger linear response region, which is not surprising considering that the system is externally detuned and off forced resonance. This linear region extends from A to B and then thereafter the planar mode saturates with increasing εP. The non-planar modes do not respond at all until point B

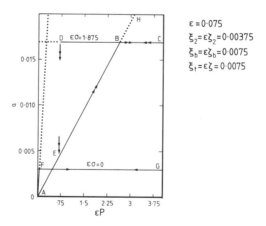

Figure 4.11 Theoretical primary response as a function of excitation parameter εP; two detuning cases $\varepsilon\sigma = 0$ and $\varepsilon\sigma = 1.875$, both cases $\varepsilon\sigma_1 = 0$ (after Bux and Roberts, 1986).

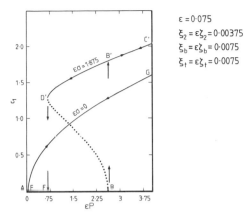

Figure 4.12 Theoretical secondary mode response (c_1) as a function of excitation parameter εP; two detuning cases $\varepsilon\sigma = 0$ and $\varepsilon\sigma = 1.875$, both cases $\varepsilon\sigma_1 = 0$ (after Bux and Roberts, 1986).

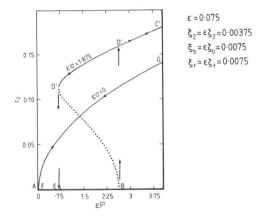

Figure 4.13 Theoretical secondary mode response (c_2) as a function of excitation parameter εP; two detuning cases $\varepsilon\sigma = 0$ and $\varepsilon\sigma = 1.875$, both cases $\varepsilon\sigma_1 = 0$ (after Bux and Roberts, 1986).

is reached. Then there is a large upward jump to B′ with a smooth increase after this as εP is increased further.

Reducing εP does not now result merely in a retracing of the planar and non-planar responses. In the planar case the nonlinear saturated solution hangs on until D where it then proceeds to suddenly jump down to E, on the linear solution line. A reduction in εP down to zero admits a linear solution which decreases proportionally. The non-planar modes retain stability of the non-zero solutions way back past the upper jump points at B′ and hold on until D′. Here the non-zero solution becomes unstable and both modes revert to the stable zero solutions at E. The dotted non-zero responses for c_1 and c_2 on Figs 4.12 and 4.13 are unstable. Hysteresis is clearly apparent (EBD on Fig. 4.11, EBB′D′ on Figs 4.12 and 4.13).

4.3.2 Four mode interactions driving simultaneous combination resonances

The system of equations in the previous example (equations (4.41), (4.44) and (4.45)) may be extended by adding one further equation which describes non-planar motion of the second bending mode. Therefore we will have a set of four equations which define three non-planar modes and one planar mode of vibration. This problem has been highlighted in the literature (Cartmell and Roberts, 1988) however we will retain the notations already defined and extend as necessary in order to admit the additional equation. To this end u_0 becomes u_{01}, ω_b becomes ω_{b1}, and we retain the general suffix for ζ (i.e. ζ_b) since for low damping cases $\zeta_{b1} \simeq \zeta_{b2} \simeq \zeta_b$. Therefore the fundamental non-planar bending equation becomes

$$\ddot{u}_{01} + \omega_{b1}{}^2 u_{01} = \varepsilon[\beta_{22}\ddot{X}_2\phi_0 - 2\zeta_b\omega_{b1}\dot{u}_{01}]. \tag{4.69}$$

The new equation for the second non-planar bending mode is

$$\ddot{u}_{02} + \omega_{b2}{}^2 u_{02} = \varepsilon[\delta_{22}\ddot{X}_2\phi_0 - 2\zeta_b\omega_{b2}\dot{u}_{02}], \tag{4.70}$$

where

$$\varepsilon\delta_{22} = \frac{m\psi_{12}}{m_0} \int_0^l (l - z)f_2{}''g\,\mathrm{d}z$$

noting also that the shape function, f, used for the linear mode in the definition for $\varepsilon\beta_{22}$, becomes f_1 for the fundamental, and f_2 for the second bending mode of the non-planar system.

The non-planar torsion equation couples also with the second non-planar bending mode to become

$$\ddot{\phi}_0 + \omega_t{}^2\phi = \varepsilon[\beta_{22}\ddot{X}_2 u_{01} + \delta_{22}\ddot{X}_2 u_{02} - 2\zeta_t\omega_t\dot{\phi}_0]. \tag{4.71}$$

Similarly the planar mode equation is extended, thus,

$$\ddot{X}_2 + \omega_2{}^2 X = \varepsilon[P\cos\Omega t + \alpha_{22}(\ddot{u}_{01}\phi_0 + 2\dot{u}_{01}\dot{\phi}_0 + u_{01}\ddot{\phi}_0)$$
$$+ \mu_{22}(\ddot{u}_{02}\phi_0 + 2\dot{u}_{02}\dot{\phi}_0 + u_{02}\ddot{\phi}_0) - 2\zeta_2\omega_2\dot{X}_2], \tag{4.72}$$

where

$$\varepsilon\mu_{22} = \frac{\psi_{22}m}{m_2} \int_0^l (l - z)f_2{}''g\,\mathrm{d}z.$$

The preceding four equations now describe the four mode interaction problem in which potential interactions between an external excitation function of frequency Ω, the second planar bending mode at ω_2, and three non-planar modes (first and second bending and first torsion), at ω_{b1}, ω_{b2} and ω_t respectively, can be explored. Certain terms which would normally appear in equations (4.69)–(4.72) containing, for example, quadratic non-linearities of the forms, $\ddot{X}_2 u_{01}$, $\ddot{X}_2 u_{02}$ and cubics of the forms $\dot{X}_2{}^2 u_{01}$, $\dot{X}_2{}^2 u_{02}$ have purposefully been neglected. This is because their presence is not likely to be contributory to the effects to be investigated in this section. For the full equations the reader is referred to the original paper (Cartmell and Roberts, 1988); however, attention is drawn to the changes in notation employed here in order to preserve some similarity with the problem of Section 4.3.1.

(a) Resonance conditions

The external resonance is as stated for the three mode problem, and is given by equation (4.47), however it is possible to extend the internal resonance case rather further by retaining the sum type combination of bending and torsion frequencies (as defined in equation (4.48)) and also considering, simultaneously, the difference resonance given below

$$\omega_2 = \omega_{b2} - \omega_t + \varepsilon\sigma_2. \tag{4.73}$$

In a physical sense this can be arranged by constructing to the mechanical details and dimensions of Fig. 4.14. It is also of interest to note that by eliminating ω_t (i.e. substituting for this from equation (4.48) in equation (4.71)) we get the following

$$\omega_2 = \tfrac{1}{2}[\omega_{b1} + \omega_{b2}] + \tfrac{1}{2}\varepsilon(\sigma_1 + \sigma_2). \tag{4.74}$$

This is similar to equation (2.204). A rather fuller treatment is given in Chapter 3, where a second-order multiple scales expansion is used with

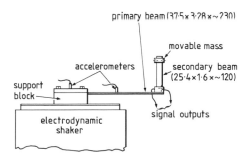

Figure 4.14 Experimental coupled beam system.

a set of three equations of motion (dealing with the fundamental and second bending modes and the fundamental torsion mode) for a base excited cantilever beam – part of this physical model. What is particularly striking about equation (4.72) is that it is close in form to equation (2.204), except that the detuning parameter is differently defined.

So we can conclude that provided the resonances defined in equations (4.48) and (4.73) are in close proximity then that of equation (4.74) will also hold.

This seems particularly significant in the light of earlier work (Sections 2.4.1(c) and 3.1.1(c)) which shows that this resonance is normally a second-order expansion effect. Here however we can show that two first-order combinations will imply its presence when they are reasonably close.

Clearly then, $\varepsilon\sigma_1$ and $\varepsilon\sigma_2$ need to be small, meaning in turn that $\frac{1}{2}\varepsilon(\sigma_1 + \sigma_2)$ will be small too, with the result that the resonance of equation (4.74) will not be too far off perfect tuning.

The additional presence of equation (4.70) and the extra terms in equations (4.71) and (4.72), as compared with the example of Section 4.3.1 results in one more slow time equation (making an autonomous set of six equations in all), and the generation of some more terms within them. These equations are

$$\omega_2 a' = -\zeta_2 \omega_2^2 a + \frac{P}{2}\sin\gamma_1 + \frac{\alpha_{22}}{4}(\omega_{b1} + \omega_t)^2 c_1 c_2 \sin\gamma_2$$

$$+ \frac{\mu_{22}}{4}(\omega_{b2} - \omega_t)^2 bc_2 \sin\psi, \tag{4.75}$$

$$\omega_2 a\gamma_1' = \omega_2\sigma a - \frac{\alpha_{22}}{4}(\omega_{b1} + \omega_t)^2 c_1 c_2 \cos\gamma_2$$

$$- \frac{\mu_{22}}{4}(\omega_{b2} - \omega_t)^2 bc_2 \cos\psi + \frac{P}{2}\cos\gamma_1, \tag{4.76}$$

$$\omega_{b1} c_1' = -\zeta_b \omega_{b1}^2 c_1 - \frac{\beta_{22}}{4}\omega_2^2 ac_2 \sin\gamma_2, \tag{4.77}$$

$$\omega_{b2} b' = -\zeta_b \omega_{b2}^2 b - \frac{\delta_{22}}{4}\omega_2^2 ac_2 \sin\psi, \tag{4.78}$$

$$\omega_t c_2' = -\zeta_t \omega_t^2 c_2 - \frac{\beta_{22}}{4}\omega_2^2 ac_1 \sin\gamma_2$$

$$+ \frac{\delta_{22}}{4}\omega_2^2 ab \sin\psi, \tag{4.79}$$

$$\gamma_2' - \psi' = \sigma_1 - \sigma_2 - \frac{\beta_{22}\omega_2^2 ac_1}{2\omega_t c_2}\cos\gamma_2$$

$$-\frac{\delta_{22}\omega_2{}^2 ab}{2\omega_t c_2}\cos\psi - \frac{\beta_{22}\omega_2{}^2 ac_2}{4\omega_{b1} c_1}\cos\gamma_2$$

$$+\frac{\delta_{22}\omega_2{}^2 ac_2}{4\omega_{b2} b}\cos\psi. \tag{4.80}$$

As in previous cases we have $a = a(T_1)$; $c_1 = c_1(T_1)$; $c_2 = c_2(T_1)$; $b = b(T_1)$; $\gamma_1 = \gamma_1(T_1)$; $\gamma_2 = \gamma_2(T_1)$; $\psi = \psi(T_1)$.

Only a rather restricted set of results have been obtained from numerical integration of these equations, mainly showing that for certain trial data the system does in fact settle into steady-state conditions and that the saturation effect is once again in evidence. Slightly earlier work by Bux and Roberts (1986) – part of which is summarized in Section 4.3.1 – also included a model of a four mode interaction problem and they too found that the resulting slow time equations would not yield meaningful solutions as functions of external detuning, $\varepsilon\sigma$ (or alternatively external frequency ratio, Ω/ω_2), when integrated numerically. The reasons for these problems are not yet clear and further research is required. For the present we can derive time responses as shown in Figs 4.15(a)–(d) and responses as functions of excitation acceleration as given in Figs 4.16(a)–(d). In the case of Figs 4.15(a)–(d) it is the envelope of the peak response that is given.

For the results given in Figs 4.15(a)–(d), it is found by inspection that the initial conditions need to be non-zero because of equation (4.76), which will contain a in the denominators of the right-hand side terms when the equation is divided through by $\omega_2 a$ to isolate $\gamma_1{}'$. Clearly zero initial conditions for a, b, c_1 and c_2 would immediately cause 'divide by

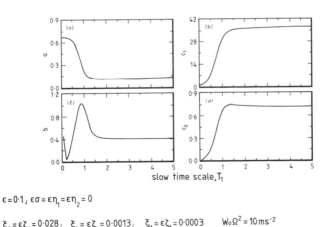

$\varepsilon = 0\cdot1$; $\varepsilon\sigma = \varepsilon\eta_1 = \varepsilon\eta_2 = 0$

$\xi_2 = \varepsilon\zeta_2 = 0\cdot028$; $\xi_b = \varepsilon\zeta_b = 0\cdot0013$; $\xi_t = \varepsilon\zeta_t = 0\cdot0003$ $W_0\Omega^2 = 10\,\text{ms}^{-2}$

Figure 4.15 Theoretical responses with slow time-scale T_1 for four mode autoparametric interaction problem.

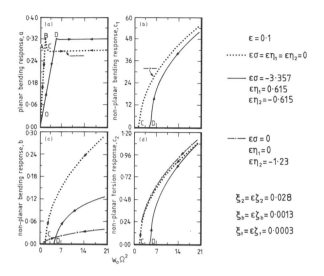

Figure 4.16 Theoretical responses as a function of support excitation acceleration for four mode autoparametric interaction problem.

zero' errors, therefore very small, but non-zero, initial conditions are chosen. There seems to be little noticeable system sensitivity to start-up conditions. Although interaction is only theoretically predicted by this model over a small range of external frequencies there is no doubt that it does occur very vigorously in certain cases of data. In the case chosen, the system displays a transient response in modes a and b (whilst modes c_1 and c_2 gradually build up) until the slow time variable has reached around $T_1 = 2.5$, after which the four modes settle down to steady state responses. Further integration over a long 'time' period verifies that the four responses are steady and constant. Although in this case mode a responds at frequency ω_2 (which, with no detuning, is exactly equal to Ω) and is therefore synchronous, modes b, c_1 and c_2 respond non-synchronously at their respective frequencies, these are well removed from Ω. Increasing the excitation magnitude, P, would result in a similar, but a more vigorously responding regime of responses with time.

In Figs 4.16(a)–(d) several time integrations are presented for fixed frequency data, but variable excitation acceleration – this being a derivable function of P, refer to equation (4.41).

The results consist of successive integration end-points taken after such a time that steady-state conditions are assured.

Three cases of external and internal tuning are quoted which in fact merge into only two cases for modes a and c_1. The saturation effect of the planar mode *a* is evident again, and the transition into nonlinear behaviour of the non-planar modes is clear at C/D and beyond. Numerical integration does not predict overhanging (i.e. hysteretical) behaviour in any of the cases.

The most significant result is that the second non-planar bending mode *b* is much more susceptible than the other to shifts in internal and external tuning. Both the dotted and solid lines define responses for which, effectively, $\frac{1}{2}\varepsilon(\sigma_1 + \sigma_2)$ in equation (4.74) reduces to zero and therefore for which we have $\omega_2 = \frac{1}{2}[\omega_{b1} + \omega_{b2}]$ exactly. The solid line (noticeably reduced c_2) is for $\sigma_1 = 0$; $\sigma_2 = 0$, therefore for detuned equations (4.48) and (4.73).

4.4 USEFUL REFERENCES AND FURTHER READING

The previous sections in this chapter have dealt with some fundamental but important problem types, however, as in Chapters 2 and 3, a wealth of important work and phenomena has not been mentioned. In an attempt to paint a slightly broader picture of the field this section summarizes the efforts and findings which have been presented by a selection of researchers over the past couple of decades. It should be noted that the amount of work that is included here is extremely limited, and exclusion from this discussion is more a reflection of the limitations of this book than anything else. Once again the reader is referred to the two major texts that treat autoparametric problems (Nayfeh and Mook, 1979; Schmidt and Tondl, 1986) for further references.

A generalized analytical treatment by Sethna of a holonomic system with two degrees of freedom, and a 2 : 1 internal ratio, provided a firm phenomenological foundation for much of the two mode interaction work that was to come (Sethna, 1965). This seminal paper demonstrated that the method of averaging could be used to solve the nonlinear differential equations of motion (with weak, quadratic, terms) and it explored the stability of the proposed system to small perturbations (i.e. to small variations of the responses about their equilibrium values). A two mass pendulum example was chosen to provide some sort of physical foundation. This is really of passing interest since the analytical presentation in this paper is of principal significance. Typical autoparametric response curves are given for different damping parameter levels.

Following on from this an important landmark in the useful application of autoparametric interaction was set by Haxton and Barr in the writing of their paper on the autoparametric vibration absorber (Haxton and Barr, 1972). Refer to Section 4.2 for further explorations of their work. Barr continued to develop the important theme of applications in a two part paper dealing with parametric and autoparametric resonance in a structure containing a liquid (Ibrahim and Barr, 1975). The problem entailed an investigation into the potential for autoparametric coupling between vertical (and horizontal) motion of a vessel partially full of liquid and the sloshing type of motion of the liquid itself. Examples of such structures abound, and include aircraft and vehicle fuel tanks, water tanks on elevated supports and space craft problems where the propellant is in liquid form.

All the problems of the preceding sections of this chapter have considered harmonic excitation functions and are called deterministic problems because of this. This means that the excitation is therefore a predictable, repetitive, function of time which invariably can be simply represented by a harmonic function. There is however, another possibility here where the excitation is not predictably repetitive (not periodic) but is random in nature. The possibility of autoparametric interactions in a structure which is excited by some form of random excitation is of great interest, and potentially of great practical significance because such functions commonly occur in nature (i.e. wind and wave loading, and atmospheric turbulence). A generalized two degree of freedom autoparametric type system (along the lines of that of Section 4.1) has been investigated (Ibrahim and Roberts, 1976) and it was shown in this work that large random and also quasi-static motions of the system could occur under the principal internal resonance condition. Several useful texts are available in this field (Crandall and Mark, 1963; Bolotin, 1984; Ibrahim, 1985), the latter dealing particularly with the random excitation of parametric systems. A further paper on this theme was published by Roberts (1980) who investigated response stability in some detail both for the generalized theoretical two degree of freedom autoparametric case and also in an experimental rig comprising two coupled cantilever beams. Both this and the preceding papers proposed a statistical method for the treatment of the governing equations because of the random nature of the applied excitation. This technique, known as Gaussian closure, is well outside the scope of this book and therefore reference should be made to the following for further details: Ibrahim and Roberts (1976), Ibrahim and Roberts (1977), Roberts (1980) and Ibrahim (1985). The paper by Roberts (1980) also presented a perturbational type of solution technique applied to the governing equations with the random applied excitation term. The result of this was a prediction of system stability which agreed with that of the Gaussian closure method but only for cases of low damping.

There is another type of response behaviour which can, on occasion, be observed in autoparametrically coupled systems; this has come to be termed chaotic response action. Chaos is usually regarded as apparently random-like response motions of the system, these being the result of a completely deterministic type of excitation however. An interesting paper on such observed behaviour in a theoretical two degree of freedom autoparametric system (similar to that investigated by Haxton and Barr) was published in 1983 (Hatwal *et al.*, part 2, 1983). They concluded that the responses enter a chaotic regime for certain combinations of forcing amplitude and frequency. For these parameter combination cases the numerical integrations employed were shown to be non-convergent. Further discussions of chaos in vibration systems are embodied in Chapter 5.

Some useful and detailed papers considering generalized structural problems in which initial element curvature and the nonlinear effect of midsurface stretching are included have been published fairly recently (Mook *et al.*, 1985, Mook *et al.*, 1986, Plaut *et al.*, 1986). These papers have shown that certain modifications to the expected autoparametric interactions can occur in these cases, and that interesting subharmonic internal resonances ($\Omega \simeq 2\omega_n, 3\omega_n$) and superharmonic internal resonances ($\Omega \simeq \omega_n/2, \omega_n/3$) come into effect.

A recent paper on four mode autoparametric interactions (Ashworth and Barr, 1987), in which the authors specifically modelled a T tail of an aircraft structure, showed that for this particular problem interaction between all four modes was difficult to achieve, however some interesting numerical results were obtained for reduced cases.

5

Phase plane concepts, and chaotic vibrations

5.1 AN INTRODUCTION TO THE PHASE PLANE

5.1.1 The undamped linear ocillator

The simple, conservative, single degree of freedom equation of motion (equation (1.10)) has been shown to have a solution which oscillates at a specific natural frequency for all time. The amplitude of this is governed by the initial conditions – those initial values of displacement and velocity at time zero. Therefore the solution to this equation is as shown in equation (1.15); both the equation of motion and the solution being restated here for convenience

$$\text{E.O.M.: } m\ddot{x} + kx = 0 \rightarrow \ddot{x} + \omega^2 x = 0; \qquad \omega^2 = k/m.$$

$$\text{Solution: } x(t) = x(0)\cos\omega t + (\dot{x}(0)/\omega)\sin\omega t.$$

The phase plane is a two-dimensional graph with displacement and velocity, respectively, on the x and y axes. Therefore we can express the simple behaviour of our conservative single degree of freedom oscillator on the phase plane by making the following transformation

$$y = \dot{x}. \qquad (5.1)$$

Substituting this into the equation of motion (1.10) and using equation (1.22), as above, leads to

$$\dot{y} = -\omega^2 x. \qquad (5.2)$$

We can rewrite this as

$$\frac{\mathrm{d}y}{\mathrm{d}x}\frac{\mathrm{d}x}{\mathrm{d}t} = -\omega^2 x \qquad (5.3)$$

noting that $\dot{x} = y$, and that differentiation with respect to x can be denoted by a prime, therefore,

$$y'y = -\omega^2 x. \tag{5.4}$$

Separation of variables gives

$$y\,dy = -\omega^2 x\,dx \tag{5.5}$$

and integration of both sides produces the following

$$\frac{y^2}{2} = \frac{-\omega^2 x^2}{2} + c, \tag{5.6}$$

where c is a constant of integration. Rearranging equation (5.6) gives the familiar type of equation for an ellipse,

$$\omega^2 x^2 + y^2 = c. \tag{5.7}$$

In order to use equation (5.7) we would have to be able to find a range of values for the independent variable x. Although x is an independent variable in this equation it is in fact completely dependent on time t, as shown above in the solution $x(t) = x(0)\cos \omega t + (\dot{x}(0)/\omega)\sin \omega t$. So by taking a set of initial conditions, such as a displacement perturbation $(x(0) = 10; \dot{x}(0) = 0)$, x may be evaluated and therefore so may y by means of equation (5.7). The relationship between the calculated time function $x(=x(t))$ and the phase plane diagram is clearly stated in

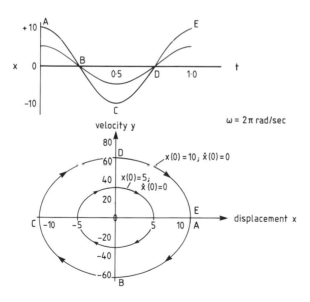

Figure 5.1 Time response and phase plane orbits for an undamped, unforced linear oscillator.

Fig. 5.1. The inner ellipse (corresponding to the lower amplitude time function) is for the case $x(0) = 5$; $\dot{x}(0) = 0$ and shows the phase plot for this altered initial state. As in the previous case the system will proceed around its defined ellipse for all time, neither gaining nor losing energy. The path of the phase plot is called the trajectory.

5.1.2 The damped linear oscillator

Having shown that the initial conditions, in the form of an initial displacement perturbation, determine the point of commencement of the resulting phase plane trajectory and therefore the size of the ellipse so defined, we now consider the effect of introduced damping. It should be noted that we are concerned with relatively light, subcritical, damping for which decaying, but oscillatory, motion will occur. Therefore the equation of motion of interest is now

$$m\ddot{x} + c\dot{x} + kx = 0 \qquad (5.8)$$

(refer to Section 1.4 and equations (1.24) and (1.25)).

Dividing through by m and introducing transformation (5.1) leads to

$$\ddot{x} + 2\xi\omega\dot{x} + \omega^2 x = 0 \qquad (5.9)$$

and therefore

$$\dot{y} = -2\xi\omega y - \omega^2 x. \qquad (5.10)$$

In order to plot this accurately it is more convenient to use a numerical integration package; so we therefore integrate equations (5.1) and (5.10) and the results for initial conditions $x(0) = 1$; $\dot{x}(0) = 0$ are as depicted in Fig. 5.2.

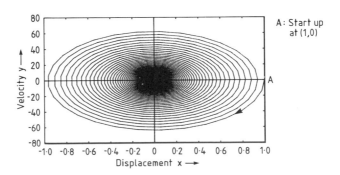

Figure 5.2 Phase portrait for a damped, unforced, linear oscillator.

The ellipse of Fig. 5.1 has here been replaced by a spiral trajectory starting at A and eventually ending up at the origin. This is commensurate with the declining transient $x(t)$ for this equation of motion. β Thus the system will always come to rest at an equilibrium state at the origin, and so this equilibrium state at $(0, 0)$ must be stable. It is called a point attractor. Changing the initial conditions has the effect of producing a larger or a smaller spiral trajectory accordingly, which will ultimately end up at the stable point attractor of the origin. The damped linear oscillator is described as being structurally stable since for any combination of damping and initial perturbation the phase plane trajectory will always spiral in towards the point attractor at the origin. Clearly for very heavy damping (i.e. above the critical) the system will respond, but in a non-oscillating manner, and the spiralling trajectory will be replaced by a node which ends up in the same point attractor at the origin.

5.1.3 The forced, damped, linear oscillator

This is described by equation (1.24) which can be rewritten in the following form by means of transformation (5.1)

$$\dot{y} = -2\xi\omega y - \omega^2 x + \frac{F_0}{m}\cos\Omega t. \tag{5.11}$$

Numerical integration of equations (5.1) and (5.11) leads to the phase portrait of Fig. 5.3. It is interesting to note that both the complementary function and the particular integral are evident as a result of, first, the initial starting-up transient (C.F.), and then the ensuing steady state response-described by the P.I. This is due to energy being supplied by the excitation and being dissipated through the damping term.

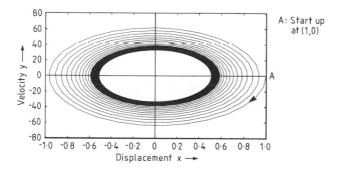

Figure 5.3 (a) Phase portrait for a damped, forced, linear oscillator: $(x(0) = 1, \dot{x}(0) = 0)$;

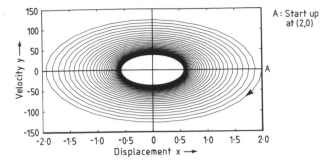

Figure 5.3 (b) Phase portrait for a damped, forced, linear oscillator: $(x(0) = 2, \dot{x}(0) = 0)$.

The P.I. part of the solution looks like a flattened ellipse on the phase portrait of Fig. 5.3(a) and is called a periodic attractor. Different initial conditions will result in different C.F. defined transients which will describe different spiralling trajectories in the phase plane. The important point is that such trajectories will always converge to the same periodic attractor (provided the excitation and damping remain the same throughout). The periodic attractor represents a limit cycle in phase space. A case for different initial conditions is given in Fig. 5.3(b).

5.1.4 The forced, undamped, linear oscillator

In this case we are putting energy into the oscillator, but have removed the dissipation term. Thus the governing equation is simply

$$\dot{y} = -\omega^2 x + \frac{F_0}{m}\cos\Omega t \qquad (\text{where } \Omega = \omega). \qquad (5.12)$$

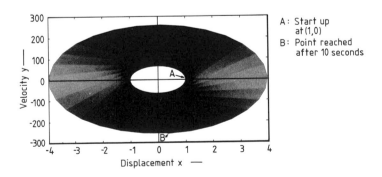

Figure 5.4 Phase portrait for an undamped, forced, linear oscillator.

Integrating this in conjunction with equation (5.1) allows us to plot the phase portrait of Fig. 5.4 which commences from initial conditions of $(1, 0)$. As there is no damping at all the response grows with time, and therefore the trajectory spirals out from the starting point. The only limit on the spiral growth from the initial conditions is the time over which it is allowed to develop.

5.1.5 The forced oscillator with linear damping and nonlinear restoring force

Here we take a form of equation (3.123) in which the cubic term is positive (hardening spring) and rewrite it such that we have the following,

$$\dot{y} = -2\xi\omega y - \omega^2 x - hx^3 + \frac{F_0}{m}\cos\Omega t. \tag{5.13}$$

In Fig. 3.10 the multivalued solution phenomenon so typical of systems modelled by equations such as (3.123), is quite apparent, and within the approximate region $1.22 < \Omega/\omega_1 < 1.51$ we see that three solutions are in fact potentially realizable. As has been stated, the middle solution of the three is unstable and therefore would not appear as a result of the numerical integration of equations (5.1) and (5.13). However, numerical integration could converge on either the upper or the lower solution; the actual one being determined by the chosen initial conditions. The obvious outcome of this for a phase plane trajectory is that two limit

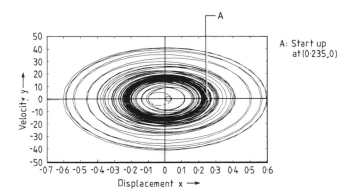

Figure 5.5 Phase portrait for a forced oscillator with linear damping and nonlinear restoring force with start up at the lower solution $(x(0) = 0.235, \quad \dot{x}(0) = 0, \quad P_0 = 200, \quad \xi = 0.01, \quad \omega = 65, \quad \Omega = 71.5, \quad \Omega = 1.1\omega)$.

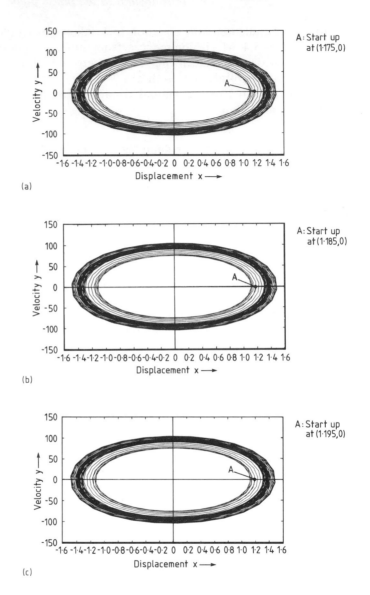

Figure 5.6 Phase portrait for a forced oscillator with linear damping and nonlinear restoring force: (a) with start up just below the unstable middle solution ($x(0) = 1.175$, $\dot{x}(0)\,0$, $P_0 = 200$, $\xi = 0.01$, $\omega = 65$, $\Omega = 71.5$, $\Omega = 1.1\omega$); (b) with start up at the unstable middle solution ($x(0) = 1.185$, $\dot{x}(0) = 0$, $P_0 = 200$, $\xi = 0.01$, $\omega = 65$, $\Omega = 71.5$, $\Omega = 1.1\omega$); (c) with start up just above the unstable middle solution ($x(0) = 1.195$, $\dot{x}(0) = 0$, $P_0 = 200$, $\xi = 0.01$, $\omega = 65$, $\Omega = 71.5$, $\Omega = 1.1\omega$).

cycles would potentially be realizable. These effects are shown in Fig. 5.5–5.7. Each figure shows a stable periodic attractor, attracting solutions which depend on the initial conditions.

Substitution of the chosen data (excitation $P_0 = 200$; damping $\xi = 0.01$; system natural frequency $\omega = 65$; excitation frequency $\Omega = 71.5$, i.e. $\Omega = 1.1\omega$) into equation (3.129) yields three real solutions when this equation is solved iteratively by computer ($A_1 = 0.235$; $A_2 = 1.185$; $A_3 = 1.375$, to 3 d.p.). Taking the lowest solution first of all, equations (5.1) and (5.13) are integrated numerically from $(0.235, 0)$. The result is as in Fig. 5.5 with the heavy elliptical ring depicting the attracting solution which passes through $(0.235, 0)$. The outer and inner spiralling trajectory denotes the C.F. part of the solution after which the clearly defined limit cycle of the periodic attractor takes over.

Figure 5.6(a) shows the phase plot for the system when started up from $(1.175, 0)$, in other words with a displacement perturbation just below the middle solution. Because the middle solution is unstable the system converges to the nearest stable solution, this being the attracting limit cycle which passes through $(1.375, 0)$. Exactly the same thing happens at start up from $(1.185, 0)$ (this being right on the middle solution) and $(1.195, 0)$ which is just above it. These cases are given in Figs 5.6(b) and 5.6(c). Since the trajectories are densely packed it is perhaps a little difficult to make out the slight differences between the three. These differences are principally in the C.F. part of the plots, after which the upper solution attracts the trajectories of all three to form stable limit cycles.

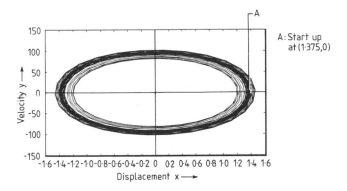

Figure 5.7 Phase portrait for a forced oscillator with linear damping and nonlinear restoring force with start up at the upper solution ($x(0) = 1.375$, $\dot{x}(0) = 0$, $P_0 = 200$, $\xi = 0.01$, $\omega = 65$, $\Omega = 71.5$, $\Omega = 1.1\omega$).

The final case is for initial conditions $(1.375, 0)$ which define the upper solution. The system quickly settles into a stable limit cycle passing through $(1.375, 0)$ after some early transient behaviour. The limit cycle is evident as the heavy elliptical portion of the trajectory, shown in Fig. 5.7.

Therefore, in conclusion, we have found two stable periodic attractors in the phase space for the non-linear (+ve cubic) system of equation (5.13) and have shown that perturbing the system at the unstable middle solution promotes steady-state behaviour at the upper solution, which means that in this case the domain of attraction of the upper solution embraces the space occupied by the unstable middle solution.

5.1.6 The linear, damped, parametric oscillator

This system is described by

$$\dot{y} = -2\xi\omega y - \omega^2 x + W\cos\Omega t.x, \tag{5.14}$$

where $\Omega = 2\omega$ for perfect tuning at principal parametric resonance. Equations (5.1) and (5.14) are numerically integrated for the following data; $\omega = 65$ ($\Omega = 130$), $\xi = 0.01$, $W = 200$, to give the results that we would expect, namely an unbounded response magnitude increasing with time. Initially there are a few decaying orbits until the system settles into its unbounded, linear response behaviour. The initially decreasing trajectory from $(1, 0)$ and then the resultant rapidly growing spiral of the trajectory with time is illustrated in Fig. 5.8.

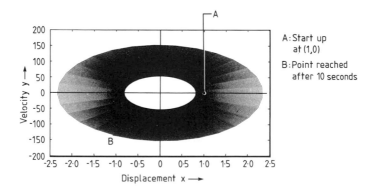

Figure 5.8 Phase portrait for a linear, damped, parametric oscillator.

5.2 A SUMMARY OF FUNDAMENTAL PHASE PLANE CONCEPTS

5.2.1 Stability

One of the simplest notions of stability applies to the case of Section 5.1.1 in which an undamped, unforced, linear, single degree of freedom oscillator is discussed. It is shown in Fig. 5.1 that such a system will respond for all time with a displacement that corresponds to the initial condition or displacement perturbation. The elliptical trajectories for the two chosen start-up conditions fit neatly one inside the other for all time because the responses are at identical natural frequencies (or the period of oscillation is identical for both perturbations). Very much closer perturbations would behave in exactly the same way with no interference to, or attraction by, one solution over the other. Such a perpetuation of an initial displacement perturbation, no matter how close to a previous one, is typical of a system such as this and therefore such motion is said to be neutrally stable.

The addition of damping radically alters the qualitative behaviour of the oscillator as clearly shown in Fig. 5.2. The point attractor located at the origin will attract all trajectories irrespective of where they start and so initial conditions are not, ultimately, preserved. The point attractor is asymptotically or structurally stable, with no topological change to the overall qualitative behaviour of the trajectory whatever the initial conditions (Thompson and Stewart, 1986). Referring back to Section 5.1.2 very heavy levels of post-critical damping such that

$$(2\xi\omega)^2 - 4(\omega^2) > 0 \tag{5.15}$$

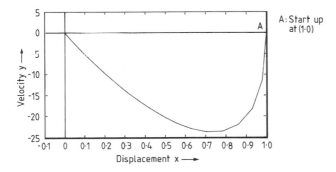

Figure 5.9 Phase portrait for oscillator of Fig. 5.2 but with post critical damping, $\xi = 1.1$; depicts a stable node.

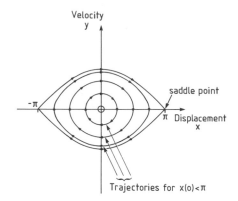

Figure 5.10 Phase portrait depicting a saddle point (unstable equilibrium) (after Thompson and Stewart, 1986).

being commensurate with real, negative, roots of the characteristic equation (refer to Section 1.4.1), prevent an oscillatory response and the trajectory of the phase plot under these conditions is as shown in Fig. 5.9. This phase plane portrait is called a node (or stable node). It should be noted that the subcritical damping case, for which the trajectory spirals into a point attractor (Section 5.1.2), has a phase portrait that is frequently described as a stable focus (Fig. 5.2). The intermediate case of critical damping, where the roots of the characteristic equation are real and equal, generates a response which is borderline between the oscillatory and the non-oscillatory. The phase portrait for this condition is very similar to that of Fig. 5.9 except that the trajectory displays rather more curvature. It is called a stable inflected node.

A further useful phase plane concept is that of the saddle point. This is usually defined as an unstable equilibrium point and conventionally the simple physical model associated with this is the swinging pendulum (Thompson and Stewart, 1986), which illustrates the point when standing vertically on its end. Figure 5.10 shows various trajectories for the undamped free oscillation case with different displacement perturbations. The saddle point is at $(\pi, 0)$ with cases increasing from below the saddle up to the singularity actually at the saddle shown in Fig. 5.10.

5.2.2 Phase plane representations using the Poincaré map technique

The simple limit cycle, descriptive of a system in steady state may be completely represented in the x, $\dot{x}$ plane for all time. However this repetitive, orbital, behaviour is not, for example, truly representative of such a system after start up and before steady state is attained. In this interval of time the system is attempting to reach a steady state, but has not yet done so, and so the trajectory in the phase space is not a limit cycle. Therefore if we were to plot the phase portrait in three dimensions including time $(x, \dot{x}, t)$ we would find that the trajectory would have the form of a three-dimensional spiral receding into the picture. Once steady state is attained the elliptical limit cycle would simply stretch out in time taking on the form of a cylindrical spiral with an elliptical cross-section.

If we now reconsider Fig. 5.5, the case of the lower solution of the Duffing type problem of Section 5.1.5, we see that there is a considerable amount of non-steady oscillation of the system below and above the attracting ring of the limit cycle which takes place before it settles into steady-state motion. The crossing over of orbits that occurs during this pre-steady-state phase obscures much of the trajectory's path. Clearly a three-dimensional picture would be more illustrative of the response of

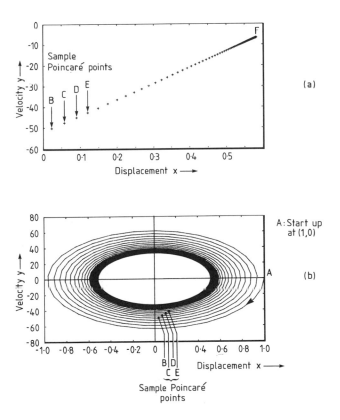

Figure 5.11 (a) Poincaré map of forced, damped, linear oscillator response; and (b) corresponding phase portrait.

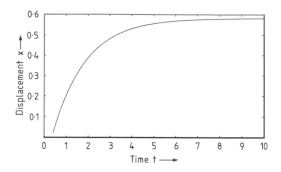

Figure 5.12 Time response for system of Fig. 5.11.

the system as it tends towards steady state. Unfortunately a three-dimensional $(x, \dot{x}, t)$ plot of the phase trajectory would still be quite complex in form, and, since it would be represented in a two-dimensional diagram, would still show confusing crossovers of the trajectory. A convenient way around this difficulty is to sample the trajectory whenever $t = nT$, where n is an integer multiple and T is the period of the system's response. Therefore we would use the following

$$t = \frac{n2\pi}{\omega}, \qquad n = 1, 2, 3. \qquad (5.16)$$

It is obvious then that for a steady-state system (ignoring for now its specific characteristics and its pre-steady-state response behaviour) such a sampled phase plot would contain one point only. This is because the orbit of the trajectory of a steady-state system is repetitive, and so sampling it at times given by equation (5.16) would produce identical $(x, \dot{x})$ points for all n.

This technique is known as Poincaré mapping and is a most powerful tool for the depiction of complicated phase plane trajectories.

Figure 5.11(b) shows the two-dimensional phase plot of Fig. 5.3(a) and four sample points defined therein. Figure 5.11(a) illustrates the Poincaré map for the problem (forced, damped, linear oscillator) and it is clear that the spiral trajectory is now shown by means of Poincaré points which progress towards the final point at F. This point defines the stable periodic attractor, or limit cycle, generated at steady state. The first few seconds leading up to steady state are shown in Fig. 5.12. Therefore a knowledge of the specifics of the dynamical system together with computerized evaluations of the Poincaré map enables one to adequately represent complex response behaviour in a two-dimensional form. The reader is strongly recommended to investigate the following for a more detailed insight; Guckenheimer and Holmes (1983) and Thompson and Stewart (1986).

5.3 CHAOS IN VIBRATING SYSTEMS

5.3.1 A summary of development in the dynamics of chaotic systems

The field of dynamical chaos is all embracing to the extent that the phenomena that have been observed over the last thirty years cannot be exclusively claimed by any one scientific group. It is an extremely important subject area because through it we are beginning to bring seemingly diverse disciplines together under various phenomenological groupings. It is also important because it provides a rigorous and academic research framework which readily admits the hitherto awkward concept of non-repeatability in systems that appear in all aspects to be completely deterministic. The early definitive work in chaos is attributable to Lorenz (1963) which explores determinism (and the lack of it) in problems of non-periodic flow. Much of Lorenz's findings were based on attempts to model meteorological problems using relatively simple equations based on fundamental physical laws. After considerable attempts to find repetitive, periodic, behaviour in these models Lorenz eventually accepted that for certain data identical results were not obtainable for repetitive runs of the system with the data unchanged. Furthermore it became clear that these sensitivities to initial conditions would invariably produce completely unpredictable, non-periodic, results given enough time. Figure 5.13 reproduces some sample results from Lorenz's early work showing the clear divergence with time of two identical solution runs. The major contribution that Lorenz made was in the observation and subsequent analysis of these effects and in introducing them as the butterfly effect, which, stated more formally, is a sensitive dependence on initial conditions.

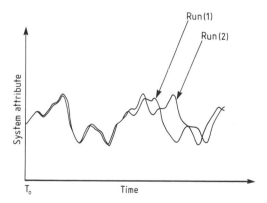

Figure 5.13 Lorenz time responses showing divergence between two runs of the system (after Gleick, 1988).

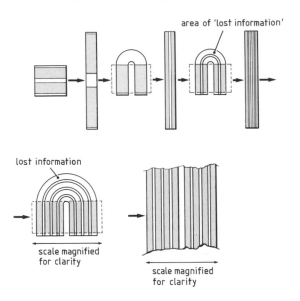

Figure 5.14 Smale horseshoe showing the effects of stretching, squeezing and folding.

At around the same time as Lorenz was discussing chaos in terms of meteorology Smale began to undertake a qualitative, topological study of the stability of general dynamical systems exhibiting chaotic behaviour (Smale, 1963). Smale's first major formalized assertion about the stability of chaotic systems was that one could not automatically assume that the presence of chaos meant that the system was unstable. In extending that proposal Smale realized that in fact a chaotically responding system could be perturbed into a different kind of response but would subsequently return to the original form of chaos once that external effect had been removed, thus showing that the stability of the system was in fact in its chaotic behaviour. The most enduring topological representation of chaotic behaviour is also attributable to Smale, and this is usually referred to as the Smale horseshoe. In itself this is a rather theoretical and abstracted proposition, but it shows that two originally close points on the structure can be shown to diverge unpredictably as its topology is changed; this provides an analogy with the more physically interpretable results of Lorenz's meteorological predictions over time. The horseshoe concept is summarized in Fig. 5.14 where the changing topology is defined by the stretching of the element with subsequent folding into the horseshoe shape. This is repeated several times with unpredictable consequences.

Some of the most fundamental work on apparently very simple

systems which can show extremely complicated behaviour was due to May who approached chaos as an observer of biological systems (May, 1976). May undertook to investigate simple nonlinear algebraic difference equations which modelled certain biological and ecological problems. Typically these were of the form

$$X_{t+1} = X_t(1 - X_t)a \tag{5.17}$$

and

$$X_{t+1} = X_t \exp[a(1 - X_t)]. \tag{5.18}$$

It was the evolution of these equations over successive iterations that was of interest. A dependency on the parameter a was observed such that stability of the system could only be preserved for certain values. For the case of equation (5.17) increasing a beyond some particular value had the effect of pushing the system into instability; stability being assessed by looking at the local slope of the plotted function at the so-called equilibrium point, where $X_{t+1} = X_t$. May's work showed that at this point of stability transition the system bifurcated (i.e. it moved

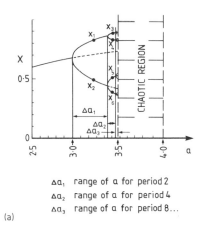

Δa_1 range of a for period 2
Δa_2 range of a for period 4
Δa_3 range of a for period 8...

(a)

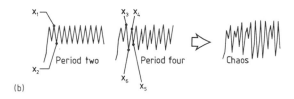

(b)

Figure 5.15 Bifurcations of X as a function of control parameter a: (a) refer to equation (5.16); (b) shows period doubling effect leading to chaos (May, 1976).

into a 'two pronged fork' type of state) so that two equilibrium points were generated, therefore

$$X_{(t+2)_1} = X_{t_1} \quad \text{and} \quad X_{(t+2)_2} = X_{t_2}.$$

Increasing a further had the effect of breaking down the stable behaviour of these two pitchfork bifurcations into further bifurcated paths. There was shown to be a very dramatic decrease in the range of a values that had to be gone through between each successive birth of bifurcations, so that this range decreased dramatically as a increased. This effect is shown in Fig. 5.15.

What May succeeded in showing with this work was that the extremely simple equation (5.17) (irrespective of its various physical foundations), demonstrated a chaotic explosion of bifurcations of the equilibrium points as a was increased, with periods from 2^0 to 2^n (refer to Fig. 5.15). It was also the case that odd-period behaviour was observed over a certain range of a. This work was summarized by May in the form of a table

Function	$X(1 - X)a$
Parameter	a
Value of parameter for which equilibrium point goes unstable	3.000
Chaotic region begins	
(point of accumulation of 'cycles' of period 2^n)	3.5700
First odd-period cycle begins	3.6786
Period 3 appears and thereafter every integer period present	3.8284
Chaotic region ends	4.000
Stable cycles found within chaos?	Yes

(Adapted from the work of May (1976))

It should be pointed out that oscillatory behaviour was, of course, at the heart of the evolution with time of equation (5.17), however the main emphasis here has been placed on the bifurcatory behaviour of X as a function of parameter a, in an attempt to show how critical the effect of this adjustment of the system is on subsequent responses. Further clarification of the concept of periodicity in the context of the bifurcations of Fig. 5.15 is given in the associated time functions on the figure.

Most writers dealing with the bifurcatory behaviour of systems modelled by equation (5.17) tend to describe the build-up to chaos as a regime of period doubling. A particularly clear and lucid account of this is given by Moon (1987) who suggests that the appearance of subharmonics (i.e. doubled, quadrupled, etc., periodic aspects to the response) can be a sure predictor of the onset of chaos. Moon gives a particularly

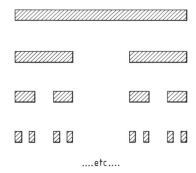

....etc....

Figure 5.16 Schematic depiction of the Cantor set.

good account of period-doubling criteria in terms of the sequence of
critical control parameter values (i.e. *a*) that control the entry of such
systems into chaotic regimes (due reference is made in Moon's work to
the originators and proponents of these ideas (Feigenbaum, 1978;
Lichtenberg and Lieberman, 1983)). The reader is recommended to look
at these references along with the following for more detailed accounts
of these effects; Guckenheimer and Holmes (1983), Thompson and
Stewart (1986) and Moon (1987).

   Another important conceptual model used in studies of chaotic
systems is the Cantor set, and its introduction to chaos is attributable to
Mandelbrot (1983). The point behind this representation is that it neatly
models the infinitely layered structure of a chaotic set of phase plane
trajectories. The Cantor set is abstract in the sense that we merely
consider a line as a base model, whose middle third is removed, and
then the remaining outer thirds are treated in the same manner and so
on as illustrated in Fig. 5.16. Mathematically the Cantor set has an
infinitely full structure reminiscent of the chaotic attractor (of a Lorenz
weather system for example) but it has zero volume. Mandelbrot's
physical hook for the role of the Cantor set centred around problems of
computer data corruption in telephone lines, and the most important
finding here was that the intermittent errors that were detected contained periods of error-free transmission, and, that the smaller error-
bursts interspersed with these error-free stretches themselves contained
error-free periods, and so on. The Cantor set contains an infinity of
points but with a total length of zero. Mandelbrot was able to propose
that the fitting of the Cantor set to a seemingly random noise problem
like this showed that such randomness in fact could be seen to contain a
form of structure. This crucial assertion underpinned his subsequent
ideas of fractal structure which were necessary for the development of
an understanding of the detailed form of chaotic regimes (the chaotic

region of Fig. 5.15(a) for example). Fractal structure is also clearly apparent after several iterations of the Smale horseshoe, the beginnings of this are evident in Fig. 5.14. It is also interesting to note that the horseshoe exhibits a degree of information loss which occurs during folding, a characteristic of the unpredictability of chaotic systems; this is shown in Fig. 5.14.

5.3.2 Chaos in a system with one degree of freedom

It has been established in previous chapters that the presence of non-linearities in an equation of motion produces certain difficulties when we consider how we might go about obtaining an analytical solution. Conventional analysis based on the assumption of linearity is inappropriate, and we turn to the various approximation techniques that have been developed for such cases. On the whole we apply such a method (multiple scales having been promoted within this book) and approximate but useful solutions are generated, and, in many cases of interest these are steady-state in nature. Therefore, the system is recognizably nonlinear, and we are quite justified in defining it as deterministic in that predictable behaviour results from a system which contains no random parameters.

However there are cases for which the system appears to be deterministic, since there are no random parameters, but which in fact exhibits non-periodic (and non-deterministic) responses. These responses are chaotic and look to be random. It has been pointed out (Tongue, 1986) that the computed power spectral density of such a system will show up a continuum of frequencies such as one might expect to find in

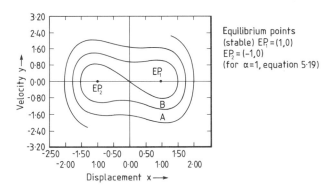

Figure 5.17 Phase portrait for the system of equation (5.19) showing nodes and saddle (after Tongue, 1986).

a random response. In addition to this, periodic response behaviour, with a finite number of frequencies present, can be re-initiated for small changes in system parameters. As Tongue states, behaviour of this sort is well outside the descriptive capabilities of nonlinear analysis; although some advances in the prediction of chaotic onset have been made (see the end of this section).

In order to study such a system we take the following equation (derived from the work of Tongue (1986))

$$\ddot{X} + \mu\dot{X} - \alpha X + \beta X^3 = \gamma\cos\Omega t. \tag{5.19}$$

Initially we investigate the unforced, free vibration, case such that we have

$$\ddot{X} + \mu\dot{X} - \alpha X + X^3 = 0, \tag{5.20}$$

where $\beta = 1$.

Numerical substitutions (such that $\mu = 0.2$, $\alpha = 1$) lead to a trajectory of the form of Fig. 5.17 where the decaying solution is attracted to the static equilibrium points. These are at $X = 0$ (saddle point unstable) and $X = \pm\sqrt{\alpha}$ (nodes), and so for start-up cases in region A these will be an attraction to the stable node at $X = \sqrt{\alpha}$, and conversely for start-up in region B, the resulting trajectory will be attracted to the stable node at $X = -\sqrt{\alpha}$. The important result of this is that the resultant motion of the system is dependent on initial conditions.

The stability (or otherwise) of the three equilibrium points is most clearly assessed by generating the potential energy function for the system and then plotting it as a function of X. To do this we take equation (5.20) (neglecting the damping term as we are only interested

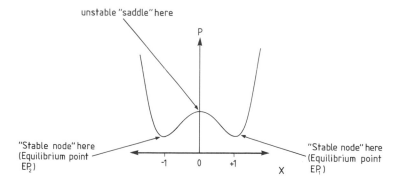

unstable "saddle" here

"Stable node" here
(Equilibrium point
EP_2)

"Stable node" here
(Equilibrium point
EP_1)

Figure 5.18 Potential energy as a function of displacement for the system of equation (5.19) showing typical 'two well potential' (after Tongue, 1986).

in kinetic and potential energies) and multiply through by $\dot{X}$, therefore leading to the following upon integration

$$\tfrac{1}{2}(\dot{X})^2 - \tfrac{1}{2}\alpha X^2 + \tfrac{1}{4}X^4 = C. \qquad (5.21)$$

Clearly the last two terms describe the system potential energy, thus,

$$P = \tfrac{1}{4}X^4 - (\alpha/2)X^2. \qquad (5.22)$$

A plot of P as a function of X is given in Fig. 5.18.

This picture helps us to visualize the system stability in the context of the three static equilibria. The wells on either side of the central hill add credence to the idea of attraction in that once the system has entered the domain or basin of attraction of the well there is no way out of it (unless the system is acted upon by a perturbation large enough to send

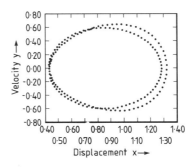

Figure 5.19 Phase portrait for period behaviour of equation (5.19) (after Tongue, 1986).

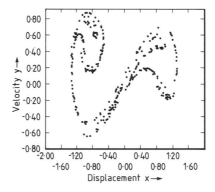

Figure 5.20 Poincaré map of chaotic attractor (system of equation (5.19)) (after Tongue, 1986).

it over the central hill into the other well) and it moves closer and closer to the equilibrium point. This is of course when damping is present.

The system can either be driven into periodicity or chaos, depending on chosen numerical data for its parameters, and, referring back to equation (5.19), the following two cases represent these regimes, respectively,

$$\ddot{x} + 0.2\dot{x} - x + x^3 = 0.3\cos(1.4t) \dots \quad \text{(Periodicity)} \quad (5.23)$$

$$\ddot{x} + 0.2\dot{x} - x + x^3 = 0.3\cos(1.29t) \dots \quad \text{(Chaos)} \quad (5.24)$$

(i.e. $\mu = 0.2$, $\alpha = 1$, $\beta = 1$, $\gamma = 0.3$ in both cases but Ω is reduced slightly in (5.24) sending the system into the chaotic regime). The phase portrait for the periodic case of equation (5.23) is given in Fig. 5.19 and is attributable to the approximate analysis of harmonic balance (Tongue, 1986). The chaotic nature of the response of the system of equation (5.24), however, is presented in Poincaré map form in Fig. 5.20. Remembering that the Poincaré map shows data at times equal to nT (see equation (5.16)) and that here the sample points are at $n2\pi/1.25$ we can identify very definite order within the chaos. This order gives a specific shape to the map for this problem and running the system over a longer time period would serve to emphasize this structure further.

The single Poincaré point at E in Fig. 5.11(a) which represents the limit cycle of Fig. 5.11(b) also qualitatively represents the Poincaré map of the system of equation (5.23). As we reduce the critical excitation frequency parameter (Ω) the response begins to bifurcate into two, four, eight cycle subharmonics (refer to Fig. 5.15). These bifurcations come in very quickly, and the Poincaré map blows up correspondingly from one point representing periodicity to (theroetically) an infinity of points representing an infinity of bifurcations (n-cycle where $n \to \infty$). The mechanism behind chaotic motions in the sort of simple nonlinear system that we have investigated here is not altogether clear cut. An intuitive reason for chaotic behaviour might be due to the imposed excitation knocking the system from one basin, or well, of attraction to another (refer to Figs 5.17 and 5.18), however, removal of the linear, negative, stiffness term of equation (5.19) removes the saddle at the origin of Fig. 5.17 (and therefore the hill at the origin of Fig. 5.18) and therefore there will be a stable equilibrium point in place of the saddle which should not, therefore, allow chaos according to the simple hypothesis given above. Unfortunately the removal of the linear term does not, in practice, kill off all chaotic regimes suggesting that a more complex behaviour mechanism than this is at the root of the chaotic response pattern of this system.

A very considerable effort has been expended in recent years in the elucidation of chaotic behaviour in Duffing type systems partly because the equation is relatively universal in its application (dynamics of a particle within a plasma, dynamics of a buckled elastic beam, and a nonlinear inductor are all examples adequately modelled by the equation) and also because it is a simple equation whose form belies the complexity of its chaotic response characteristics. For further information the reader is referred to the following: Holmes (1979), Ueda (1979, 1980), Holmes and Moon (1983).

Before leaving this introduction to chaos it is important to note that several criteria for the prediction of chaos have recently been evolved. Moon describes these in terms of empirical and theoretical predictive criteria and it is probably the latter that have most substance for the researcher in dynamics; in particular the criteria for period doubling using the so-called Feigenbaum number (Feigenbaum, 1979, 1980; Guckenheimer and Holmes, 1983), the Melnikov method (Guckenheimer and Holmes, 1983) and Chirikov's overlap criterion (Lichtenberg and Lieberman, 1983). It is also interesting to note that the onset of chaos in the Ueda form of the Duffing oscillator ($\ddot{x} + a\dot{x} + x^3 = P\cos t$) can apparently be predicted using perturbation theory for which the stability of the predicted subharmonics provides the indicator of chaos.

Finally this cursory discussion of chaotic identifiers (or of a system's potential for chaotic behaviour) would not be complete without mentioning the technique of Liapunov exponents. The Liapunov exponent is a measure of trajectory divergence in the phase space, however further development of the theory is well outside the scope of this introductory text, and the interested reader is recommended, in particular, to investigate the work of the following (Wolf, 1984; Wolf *et al.*, 1985), for full and informative accounts.

References

Abramson, H. N. (1966) *The Dynamic Behaviour of Liquids in Moving Containers*, NASA–SP–106.

Anderson, T. L. and Moody, M.L. (1969) *J. Eng. Mech. Div. (ASCE)*, **95** 665–77.

Andronov, A. A. and Chaikin, C. E. (1949) *Theory of Oscillations* (trans. from Russian), Princeton University Press, Princeton, NJ.

Arscott, F. M. (1964) *Periodic Differential Equations: An Introduction to Mathieu, Lame, and Allied Functions*, Macmillan, New York.

Ashworth, R. P. and Barr, A. D. S. (1987) *J. Sound Vib.*, **118** (1), 47–68.

Barr, A. D. S. and McWhannell, D. C. (1971) *J. Sound Vib.*, **14** (4), 491–509.

Blackwell, L. A. and Kotzebue, K. L. (1961) *Semiconductor–Diode Parametric Amplifiers*, Prentice-Hall, Englewood Cliffs, NJ.

Bogoliubov, N. N. and Mitropolsky, Y. A. (1961) *Asymptotic Methods in the Theory of Non-Linear Oscillations* (trans. from Russian), Gordon and Breach, New York.

Bolotin, V. V. (1964) *The Dynamic Stability of Elastic Systems*, Holden-Day, San Francisco.

Bolotin, V. V. (1984) *Random Vibration of Elastic Systems*, Mechanics of Elastic Stability Series, Nijhoff, The Hague.

Bux, S. L. and Roberts, J. W. (1986) *J. Sound Vib.*, **104** (3), 497–520.

Cartmell, M. P. (1984) PhD thesis, University of Edinburgh, *Combination Instabilities and Non-Linear Vibratory Interactions in Beam Systems*.

Cartmell, M. P. and Roberts, J. W. (1987) *Strain*, **33** (3), 117–26.

Cartmell, M. P. and Roberts, J. W. (1988) *J. Sound Vib.*, **123**, (1), 81–101.

Caughey, T. K. (1960) *J. Appl. Mech.*, **27**, 269–71.

Crandall, S. H. and Mark, W. D. (1963) *Random Vibration in Mechanical Systems*, Academic Press, New York.

Dalzell, J. F. (1967) *Exploratory studies of liquid behaviour in randomly excited tanks: Longitudinal excitation*, Southwest Research Institute, San Antonio, Tech. Report no. 1.

Den Hartog, J. P. (1934) *Mechanical Vibrations*, McGraw-Hill, New York.

Den Hartog, J. P. (1956) *Mechanical Vibrations*, McGraw-Hill, New York.

Dodge, F. T., Kana, D. D. and Abramson, H. N. (1965) *A.I.I.A.J.*, **3**, 685–95.

Dugundji, J. and Mukhopadhyay, V. (1973) *J. Appl. Mech.*, **95** (3), 693–8.

Evans-Iwanowski, R. M. (1976) *Resonance Oscillations in Mechanical Systems*, Elsevier, Amsterdam.

Feigenbaum, M. J. (1978) *J. Stat. Phys.*, **19** (1), 25–52.

Feigenbaum, M. J. (1979) *Phys. Lett.*, **74A**, 375–9.

Feigenbaum, M. J. (1980) *Los Alamos Sci.* (summer), 4–27.

Floquet, G. (1883) *Ann. Sci. Ecole Norm. Sup.*, **12**, 47–89.

Gershuni, G. Z. and Zhukhovitskii, E. M. (1963) *PMM*, **27** (5), 779–83.

Guckenheimer, J. and Holmes, P. J. (1983) *Nonlinear Oscillations, Dynamical Systems, and Bifurcations of Vector Fields*, Springer-Verlag, New York.

Guttalu, R. S. and Hsu, C. S. (1984) *J. Sound Vib.*, **97** (3), 399–427.

Handoo, K. L. and Sundararajan, V. (1971) *J. Sound Vib.*, **18** (1), 45–53.

HaQuang, N. and Mook, D. T. (1987a) *J. Sound Vib.*, **118** (2), 291–306.

HaQuang, N. and Mook, D. T. (1987b) *J. Sound Vib.*, **118** (3), 425–39.

Hatwal, H., Mallik, A. K. and Ghosh, A. (1983) *J. Appl. Mech.*, **50**, parts 1 and 2, 657–68.

Haxton, R. S. and Barr, A. D. S. (1972) *J. Engng. Ind.*, **94** (1), 119–24.

Hayashi, C. (1964) *Nonlinear Oscillations in Physical Systems*, McGraw-Hill, New York.

Hill, G. W. (1886) *Acta Math.*, **8**, 1–36.

Holmes, P. J. (1979) *Phil. Trans. Roy. Soc. London, A.* **292** 419–48.

Holmes, P. J. and Moon, F. C. (1983) *J. Appl. Mech.*, **50**, 1021–32.

Hsu, C. S. (1975a) *J. Sound Vib.*, **39** (3), 305–16.

Hsu, C. S. (1975b) *J. Appl. Mech.*, **42**, 176–82.

Ibrahim, R. A. (1972) *Liquid Sloshing and Related Topics*, University of Edinburgh Research Report, part 1: Bibliography.

Ibrahim, R. A. (1978) *Shock and Vibration Digest*, **10** (3), 41–47, and **10** (4), 19–47.

Ibrahim, R. A. (1985) *Parametric Random Vibration*, Research Studies Press, Wiley, Letchworth.

Ibrahim, R. A. and Barr, A. D. S. (1975) *J. Sound Vib.*, **42** (2), 159–200.

Ibrahim, R. A. and Barr, A. D. S. (1978) *Shock and Vibration Digest*, **10** (1), 15–29, and **10** (2), 9–24.

Ibrahim, R. A. and Roberts, J. W. (1976) *J. Sound Vib.*, **44** (3), 335–48.

Ibrahim, R. A. and Roberts, J. W. (1977) *Z. Angew. Math. Mech.*, **57**, 643–9.

Ibrahim, R. A. and Soundararajan, A. (1983) *J. Sound Vib*, **91** (1), 119–34.

Jaluria, Y. (1980) *Natural Convection: Heat and Mass Transfer*, Pergamon, Oxford.

Jordan, D. W. and Smith, P. (1979) *Nonlinear Ordinary Differential Equations*, Oxford University Press, Oxford.

Kaliski, S. (1968) *Proc. Vib. Problems (Warsaw)*, **9** (1), 79–88.

Lasalle, J. P. and Lefshetz, S. (1961) *Stability by Liapunov's Direct Method with Applications*, Academic Press, New York.

Lazan, B. J. (1968) *Damping of Materials and Members in Structural Mechanics*, Pergamon, Oxford.

Lichtenberg, A. J. and Lieberman, M. A. (1983) *Regular and Stochastic Motion*, Springer-Verlag, New York.

Lorenz, E. N. (1963) *J. Atmos. Sci.*, **20**, 130–41.

Love, A. E. H. (1944) *Treatise on the Mathematical Theory of Elasticity*, First American Press, New York.

McLachlan, N. W. (1955) *Bessel Functions for Engineers*, Clarendom Press, Oxford.

Mahalingam, S. (1957) *British J. Appl. Phys.*, **8**, 145–8.

Mandelbrot, B. (1983) *The Fractal Geometry of Nature*, W. H. Freeman, San Francisco.

May, R. M. (1976) *Nature*, **261**, 459–67.

Meirovitch, L. (1986) *Elements of Vibration Analysis*, McGraw-Hill, New York.

Minorsky, N. (1962) *Nonlinear Oscillations*, Van Nostrand, Princeton.

Mitchell, R. R. (1968) *Stochastic Stability of the Liquid Free Surface in Vertically Excited Cylinders*, NASA-CR-98009.

Miyasar, A. M. and Barr, A. D. S. (1988) *J. Sound Vib.*, **124** (1), 79–89.

Mook, D. T., Plaut, R. H. and HaQuang, N. (1985) *J. Sound Vib.*, **102** (4), 473–92.

Mook, D. T., HaQuang, N. and Plaut, R. H. (1986) *J. Sound Vib.*, **107** (2), 309–19.

Moon, F. C. (1987) *Chaotic Vibrations*, Wiley, New York.

Moon, F. C. and Pao, Y. H. (1969) *J. Appl. Mech.*, **36**, 92–100.

Mote, C. D. (1968) *J. Appl. Mech.*, **90** (3), 171–2.

Naguleswaran, S. and Williams, C. J. H. (1967) *Int. J. Mech. Sci.*, **10**, 239–50.

Nashif, A. D., Jones, D. I. G. and Henderson, J. P. (1985) *Vibration Damping*, Wiley, New York.

Nayfeh, A.H. (1973) *Perturbation Methods*, Wiley, New York.

Nayfeh, A.H. (1983) *J. Sound Vib.*, **90** (2), 237–44.

Nayfeh, A. H. (1987) *J. Sound Vib.*, **119** (1), 95–109.

Nayfeh, A. H. and Mook, D. T. (1979) *Nonlinear Oscillations*, Wiley, New York.

Nayfeh, A. H. and Zavodney, L. D. (1986) *J. Sound Vib.*, **107** (2), 329–50.

Ness, D. J. (1971) *J. Appl. Mech.*, **93** (3), 585–90.

Othman, A. M. (1986) PhD thesis, University of Dundee, *An Oscillator under High Frequency Multicomponent Parametric Excitation*.

Othman, A. M., Watt, D. and Barr, A. D. S. (1987) *J. Sound Vib.*, **112** (2), 249–59.

Plaut, R. H., HaQuang, N. and Mook, D. T. (1986) *J. Sound Vib.*, **106** (3), 361–76.

Rao, S. S. (1984) *Mechanical Vibrations*, Addison-Wesley, Reading, MA.

Rayleigh, Baron (Strutt, J. W.) (1945) *The Theory of Sound*, Dover, New York.

Rhodes, J. E. (1970) *J. Appl. Mech.*, **92** (3), 1055–60.

Roberts, J. W. (1980) *J. Sound Vib.*, **69** (1), 101–16.

Roberts, J. W. (1984) *Int. J. Mech. Engng. Education*, **13** (1), 55–75.

Roberts, J. W. and Cartmell, M. P. (1984) *Strain*, **20** (3), 123–31.

Schmidt, G. and Tondl, A. (1986) *Nonlinear Vibrations*, Cambridge University Press, Cambridge.

Sethna, P. R. (1965) *J. Appl. Mech.*, **32**, 576–82.

Smale, S. (1963) Diffeomorphisms with many periodic points in *Differential and Combinatorial Topology* (ed. S. S. Cairns), Princeton University Press, Princeton, NJ, pp. 63–80.

Struble, R. A. (1962) *Nonlinear Differential Equations*, McGraw-Hill, New York.

Thomson, W. T. (1981) *Theory of Vibration with Applications*, George, Allen and Unwin, London.

Thompson, J. M. T. and Stewart, H. B. (1986) *Nonlinear Dynamics and Chaos*, Wiley, Chichester.

Timoshenko, S. P., Young, D. H. and Weaver, W. (1974) *Vibration Problems in Engineering*, Wiley, New York.

Tondl, A. (1965) *Some Problems in Rotordynamics*, Chapman and Hall, London.

Tongue, B. H. (1986) *J. Sound Vib.*, **110** (1), 69–78.

Tse, F. S., Morse, I. E. and Hinkle, R. T. (1978) *Mechanical Vibrations: Theory and Applications*, Allyn and Bacon, Boston.

Tso, W. K. and Caughey, T. K. (1965) *J. Appl. Mech.*, **32**, 899–902.

Ueda, Y. (1979) *J. Stat. Phys.*, **20**, 181–96.

Ueda, Y. (1980) Steady motions exhibited by Duffing's equation: a picture book of regular and chaotic motions, in *New Approaches to Nonlinear Problems in Dynamics* (ed. P. J. Holmes), SIAM, Philadelphia, pp. 311–22; and Explosions of strange attractors exhibited by Duffing's equation in *Nonlinear Dynamics* (ed. R. H. G. Helleman), New York Academy of Sciences, New York, pp. 422–34.

Warburton, G. B. (1976) *The Dynamical Behaviour of Structures*, Pergamon, Oxford.

Watt, D. and Barr, A. D. S. (1983) *J. Vibration, Acoustics, Stress and Reliability, and Design*, **105**, 326–31.

Wolf, A. (1984) *Quantifying Chaos with Liapunov Exponents*, Nonlin. Sci. Theory App., Manchester University Press, Manchester.

Wolf, A., Swinney, H. L. and Vasano, J. (1985) *Physica*, **16D**, 285–317.

Woodward, J. H. (1966) PhD thesis, Georgia Inst. of Tech., *Fluid Motion in a Circular tank of Sector-Annular Cross-Section when Subjected to Longitudinal Excitation*.

Wu, W. Z. and Mote, C. D. (1986) *J. Sound Vib.*, **110** (1), 27–39.

Yamamoto, T. and Saito, A. (1970) *Mem. Fac. Engng., Nagoya Univ.*, **22** (1), 54–123.

Zavodney, L. D. and Nayfeh, A. H. (1988) *J. Sound Vib.*, **120** (1), 63–93.

Index